Fiber (

Applications to Commercial, Industrial, Military, and Space Optical Systems

CHAPTER 1

Performance Requirements for Fiber Core, Cladding, and Jacket

Cutting-edge fiber optic (FO) technology brings new dimensions for optical networking, long-haul optical communications, signal processing, medical diagnosis, battlefield applications, and space surveillance and reconnaissance. Performance requirements for the fiber core, cladding, jacket, buffer, and coating will vary from application to application. Critical electrical, mechanical, thermal, and optical properties of the FO elements will be summarized as a function of wavelength and operating environmental parameters such as temperature, vibration, shock, and gravitational force. FO technology has potential applications to electro-optic and photonic devices, opto-electronic components, medical diagnosis and bio-engineering, missile guidance, phased array radars, satellite communications, long-haul telecommunication communications and data transmission, data links for classified missions, secured communications, underwater tracking and detection systems, programmable delay lines for electronic countermeasures (ECM) systems, towed-decoy delivery systems, space-based sensors, and submarine-based towed sensors.

Studies performed by the author [1] indicate that FO-based cables provide high data-transmission rates, multiple voice and video channels with minimum cross-talk, and secured communications over wideband and large distances, while operating under harsh environments. These studies further indicate that FO cables are capable of operating under nuclear radiation, electromagnetic interference (EMI), and severe thermal and mechanical operating conditions, while maintaining high mechanical integrity, stable optical performance, and improved thermal performance. Requirements for various elements of the FO cable will be defined to meet the performance requirements while under harsh environments such as high jet-engine exhaust temperatures, toxic zones, and the high gravitational forces encountered under severe uncontrollable aerodynamic conditions.

1.1 Materials for Various Elements of an Optical Fiber

Optical coplanar waveguide (CPW) or FO lines can be used for transmitting and distributing optical signals with minimum loss, low dispersion, and high reliability under nuclear, EMI, and harsh operating environments. Optical waveguides or fibers are made from glass materials such as AgCl, KCl, silicate glass, fluoride glass, chalcogenide glass, and oxynitride material [2]. A commercially available FO cable is composed of three elements: core, cladding, and shield or jacket. However, FO cables designed to operate under harsh thermal, chemical, toxic, and mechanical environments require a special coating layer, strength member, buffer layer, and inner and outer jackets. The core is made of specified refractive-index material that is surrounded by a concentric cladding material with a slightly higher refractive index. The coating material of a specified thickness is intended to protect the optical core under severe thermal and mechanical environments. The core is generally made from silica glass with an index-modifying dopant such as GeO_2. A protective coating of cushioning material such as acrylate is used to reduce the crosstalk between the adjacent fibers and the microbending that normally occurs under severe mechanical environments. For greater environmental protection and highly stable optical performance, cladding and buffer layers are surrounded by a jacket made from steel fibrous material and Kevlar strands, as shown in Figure 1–1.

Optical fiber geometry and material composition determine the discrete set of electromagnetic (EM) fields that can propagate in the fiber with minimum loss and dispersion. Both radiation modes and guided modes are present in optical fibers. Radiation modes carry EM energy out of the core so that energy is dissipated in the cladding layer. Guided modes are confined to the cores that propagate optical energy along the fiber axis, transporting information and power through the fiber. If the fiber core is large enough, it can support many guided modes. The two lowest-order guided modes are present in a circularly symmetric fiber core. When light is launched into the fiber, various modes are excited. Some light is absorbed in the jacket, and the rest is propagated into the fiber by the internal reflections between the core and the cladding (see Figure 1–1). Material requirements for the core, cladding, buffer, and jacket will be defined with particular emphasis on insertion loss, dispersion, microbending, and the power-carrying capabilities of the core and cladding.

1.2 Material Requirements for the Core

The material properties and the dimensional parameters for the core are critical because they determine the suitability for a single-mode (SM) fiber or a multi-mode (MM) fiber for a specific application. The dimensional parameters for he core and cladding define the refractive-index profiles (see Figure 1–2) for step-index MM fibers or graded-index fibers, step-index SM fibers or dispersion-shifted SM fibers.

Cores can be made from conventional glass or oxynitride glass materials. Conventional glass materials (see Figure 1–3) include silicate glass, fluoride glass, KCL glass, chalcogenide glass, and zinc chloride glass. These materials are widely used for infrared (IR) optical fibers because of lower insertion and bulk absorption losses as shown in Figure 1–3.

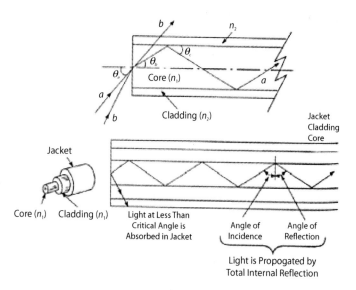

Figure 1–1 *Critical elements of an optical fiber.*

Core materials must have low values of dielectric constant and loss tangent if minimum insertion and absorption losses are the principal requirements. As it will be pointed out later, low insertion loss and material dispersion are critical for optical fibers used in long-haul telecommunications, wavelength division multiplexing (WDM) and dense wavelength division multiplexing (DWDM) systems. High-lead glasses with Pb^{+3} ions are widely used for high-performance optical-transmission lines. Optical fibers made from alkali-free aluminosilicate glasses are best suited for space applications. Studies performed by the author [1] indicate that properties of glass can be altered substantially by introducing certain materials in the core, such as Al_2O_3 to improve the mechanical properties of the core, PbO to improve the optical quality, Be_2O_2 to enhance the thermal performance, and Si_3N_4 (silicon nitride) to improve the thermal shock capability of the fiber. Absorption loss in an optical fiber can be controlled by introducing certain ions in the glass, as illustrated in Figure 1–4. Pure silica-based cores are most attractive for certain military and space applications, where high mechanical strength, improved reliability, excellent thermal-shock resistance, and chemical inertness are necessary to meet the principal requirements for high reliability and mechanical strength at elevated temperatures.

1.2.1 Mechanical Requirements for Core Materials

In military and space applications, high mechanical integrity, improved reliability, and stable optical performance are of paramount importance. Fiber cores

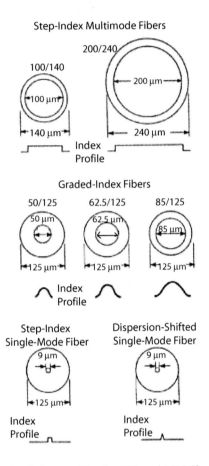

Figure 1–2 *Refractive-index profiles for SM and MM fibers as a function of core and cladding diameters.*

with high tensile strength, improved fatigue resistance, high stiffness, and excellent thermal-shock resistance must be deployed where reliability and mission success under harsh environments are the principal requirements. Cores made from cordierite glass or oxynitride glass are best suited for applications where reliability and stable optical performance are critical under severe thermal and mechanical environments. Mechanical properties of potential cores are summarized in Table 1–1.

Materials for cores, cladding, and coating with low values of coefficient of thermal expansion (CTE) should be selected to maintain high mechanical integrity and stable optical performance with minimum crushing, kinking, or dispersion.

1.2 Material Requirements for the Core 7

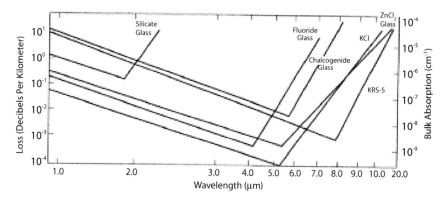

Figure 1-3 *Insertion loss and bulk absorption coefficient for various optical fiber materials as a function of operating wavelength.*

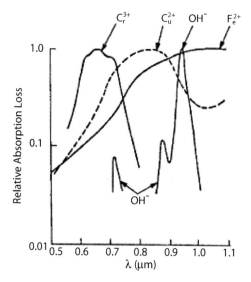

Figure 1-4 *Relative absorption loss for ions present in a silica glass as a function of operating wavelength.*

1.2.2 Requirements for Cladding

Requirements for cladding are not severe compared to those for the core. The cladding must be from a low-loss glass such as pure silica, bonded hard polymer or fluorine-doped synthetic-fused silica with desirable refractive index values to provide optimum profile. Both the dimensions and refractive index of the

Table 1–1 *Mechanical properties for potential glass-based cores [1].*

Properties	E-glass	Fused Silica Glass	Cordierite Glass	Oxynitride Glass
Tensile modulus (psi) (10^6)	10	10.8	12.6	25
Tensile strength (psi)	600,000	750,000	860,000	980,000
KNOOP hardness	460	685	715	835
Fatigue resistance	Fair	Good	Very good	Excellent
Coefficient of thermal expansion (10^{-6}) in/in/°C	9.9	0.57	4.50	2.82
Estimated cost ($/meter)	0.60	3.12	8.25	14.55

cladding layer must provide minimum scattering losses as well as an optimum refractive-index profile. A coating of appropriate material and thickness is provided on the cladding layer to achieve high-polarization preservation capability. Furthermore, materials for the cladding layer must have lower values of refractive indices to meet high coupling-efficiency requirements. Cladding-layer thickness along with a polarization preservation coating must provide optimum refractive-index profiles for both the step-index and graded-index SM and MM fibers, as illustrated in Figure 1–2.

1.2.2.1 Refractive-Index Profiles

Lower dispersion and high optical stability are dependent on the refractive-index profile of the fiber used. The core and cladding dimensions determine the refractive-index profile regardless of the fiber types, which include step-index SM fibers, step-index MM fibers, graded-index fibers and dispersion-shifted (DS) fibers as shown in Figure 1–2. For step-index SM fibers and DS fibers, the cladding diameter is significantly larger than the core diameter. However, in the case of step-index MM fibers or graded-index fibers, the cladding diameter is slightly more than the core diameter as illustrated in Figure 1–2.

Total internal reflections between the core and cladding determine the maximum acceptance angle, which is defined as

$$\sin \theta_a = [n_2/n_1] \qquad 1.1$$

where n_2 and n_1 are the refractive index for cladding and core materials, respectively, and θ_a is the acceptance angle.

Another important performance parameter is the numerical aperture (NA), which is dependent on the refractive index of the core and cladding materials. This parameter can be written as

$$NA = [\sqrt{n_1^2 - n_2^2}] \qquad 1.2$$

Higher values of the NA parameter (typical values are between 0.25 to 0.37) are necessary to achieve higher light-coupling efficiencies. Equation 1.2 indicates that cladding materials with lower values of refractive index (n_2) are required to achieve high coupling efficiency or low coupling loss.

1.3 Material Requirements for Coating

The bare optical fiber must be provided with a polyimide or acrylate coating of appropriate thickness. The polyimide coating thickness typically varies between 150 to 175 microns, whereas the preferred acrylate coating is close to 250 microns. The coating thickness varies from 0.0005 to 0.005 inch depending on the application and operating environments. Acrylate coating has a dielectric constant between 3.4 and 3.8, a loss tangent of less than 0.003, a maximum service temperature of 500°C and a tensile strength of better than 25,000 psi. Its low frictional coefficient of 0.17 minimizes the surface abrasion. This coating material provides one of the strongest protecting films available; offers excellent electrical properties; and allows resistance to moisture, humidity, abrasion, acids, grease, oils, and radiation (better than 10^6 rad.). In applications where hermetic sealing is desired, a thin carbon coating is required on the cladding surface. In summary, coating provides protection from chemical agents, thermally induced attenuation and phase variations, and physical damage to the surface. The material and thickness of typical coatings used by high performance single mode and multimode optical cables are summarized in Table 1–2.

1.4 Material Requirements for Buffer

A buffer layer is provided over the cladding to protect the optical fiber from excessive bending, pinching, or crushing under severe mechanical environments. In addition, the buffer, when made from expanded polytetrafluoroethylene (PTFE), significantly reduces the thermally induced attenuation and phase variations over a wide temperature range from −55°C to +150°C. High attenuation due to optical shrinkage after thermal cycling can be expected and can be minimized using a suitable material for the buffer. An SM fiber is more sensitive to bending losses compared to an MM fiber, and, hence, requirements for an SM

Table 1–2 Coating material and thickness used by ruggedized optical cables [3].

Optical Core Type (SM or MM)	Core/Cladding/ Coating Thickness (microns)	Coating Material	Wavelength (nm)
Single-mode (SM)	9.3/125/155	Polyimide	1310
Multimode (MM) graded-index	62.5/125/155	Polyimide	1310
MM graded-index	62.5/125/250	Acrylate	1310
Single-mode	9.3/125/250	Acrylate	1310
Single-mode*	9.3/125/250	Polyimide	1310
MM graded-index*	100/140/172	Polyimide	1310
Single-mode	7.5/125/245	Acrylate	1550
Single-mode	7.0/125/250	Acrylate	980
Single-mode*	7.0/125/250	Acrylate	980

* The optical cables with carbon coating on the cladding layer to provide the hermetic-sealing capability for certain applications.

fiber are different from those for an MM fiber. Studies performed by the author indicate that in addition to thermal effects, fiber routing can induce bending losses if the fiber is not adequately buffered and the outer layers pinch or crush the optical fiber. A buffer layer offers resistance to bending, crushing, kinking, thermal shock, and vibration under adverse operating environments. In summary, the PTFE buffer layer protects the optical fiber from physical damage or deformation and helps to minimize optical losses under severe thermal and mechanical conditions.

The buffer material must possess unique electrical and mechanical properties, and PTFE offers such properties. The PTFE has a low dielectric constant and loss tangent over a wide temperature range. This material does not absorb water and is unaffected by most chemicals. The buffer layer thickness can be as much as 0.10 inches or 250 microns. The tensile strength of this material varies from 2100 to 4500 psi. PTFE provides high thermal stability and excellent resistance to acids, grease, oils, and organic solvents. In addition, the PTFE buffer offers corrosion protection from soil, fungus, toxic gases, and other microorganisms.

1.5 Material Requirements for Jacket

Jackets provide protection to the optical fiber from physical damage due to crushing, bending, or kinking under harsh thermal and mechanical operating environments. For most communications optical fibers, one jacket is adequate to maintain stable optical performance and to protect the fiber from thermal shock and gravitational forces. Several jacket materials are available, including fluoropolymer, Kevlar aramid, fluorinated ethylene propylene (FEP), polyamide, Kevlar strands, Tefzel, and nylon. In summary, ruggedized versions of optical cable deploy strength members in addition to inner and outer jackets to improve reliability and provide high mechanical integrity, enhanced damage threshold levels, and stable optical performance under severe thermal and mechanical environments [3]. The ruggedized optical cables are best suited for military and space applications involving high power laser delivery systems.

The strength member (Figure 1-5) is comprised of Kevlar strands that provide the optical cable extra strength to achieve stable optical performance when the cable is subjected to vibration, mechanical shock, gravitational forces, and thermal stress. Kevlar is widely used in aromatic polyamide fibers because of its low density, high tensile strength, and extreme toughness. Kevlar jacket material offers excellent ballistic protection, high chemical resistance, remarkable tensile strength and improved resistance to radiation. No other jacket material can beat or meet the performance capability of Kevlar in terms of ballistic protection and mechanical integrity. Studies performed by the author indicate that this material offers a tensile strength greater than 525,000 psi, a tear strength of 150,000 psi, and an extremely low coefficient of thermal expansion that is close to zero. Kevlar is best suited for both the inner and outer jacket because of its outstanding mechanical properties, excellent chemical resistance, and remarkable thermal stability.

The inner jacket can be made from a fluoropolymer material that offers added strength and isolation with minimum cost. The inner jacket surrounds the buffer layer as illustrated in Figure 1-5. The inner jacket thickness varies from 400 to 850 microns depending upon the protection required against crushing, kinking, or thermal instability. Additional protection or performance improvement in these areas can be incorporated in the design of the outer jacket. Excessive jacket thickness will increase the weight, size, and cost of the optical cable. The inner jacket sometimes is surrounded by a strength member or layer made from Kevlar strands. This strength member provides additional strength and isolation under severe thermal or mechanical stress environments. The thickness of this strength layer varies from 250 to 450 microns depending upon the additional strength requirements and weight and size constraints.

The outer jacket is needed for highly ruggedized optical cables where exceptional high reliability and stable optical performance are critical under harsh thermal and mechanical environments. The outer jacket can be made from

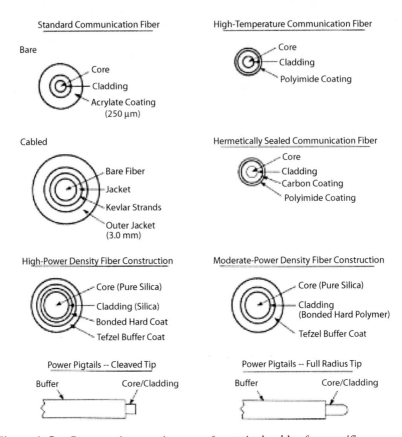

Figure 1–5 *Construction requirements for optical cables for specific applications.*

polyamide, Tefzel, or FEP material. The outer jacket thickness can vary from 1000 to 3000 microns depending on the thermal, mechanical, and optical performance requirements. FEP material offers excellent mechanical properties, namely, a tensile strength greater than 500,000 psi, a comprehensive strength greater than 200,000 psi, and a shear strength of more than 85,000 psi. If high hardness and stiffness are desired, one might consider using flexiglass material [4] for the outer jacket because of its excellent tensile strength, compressive strength, hoop strength, torsional strength, and CTE properties. Regardless of the material used, the outer jacket must provide increased resistance to corrosion, radiation, EMI, chemicals, toxic gases, crushing, kinking, puncture, bending, and thermal shock under severe thermal and mechanical operating environments.

1.6 Various Loss Mechanisms in Optical Fiber

Comprehensive studies undertaken in the early 1960s and 1970s by well-known scientists indicate that fused silica glass fibers could play a key role in long distance telecommunications systems and high data transmission. Studies performed in 1968 by Standard Telecommunications Laboratories in England revealed that glass fibers with the lowest impurities would be best suited for long-haul communications systems [5]. At that time, typical glass fiber losses were close to 1000 dB/km. However, subsequent research and development activities undertaken by Bell Labs, Corning Glass Works, Nippon Sheet Glass, German AEG-Telefunken, and Simens brought significant reductions in the SM fiber loss—under 20 dB/km by 1975, 5dB/km by 1985, 0.5 dB/km by 1995, and less than 0.35 dB/km by 2000.

Various loss mechanisms introduce losses in an optical fiber. These loss mechanisms include material absorption, material scattering including active and/or passive components, internal scattering between the core and cladding, radiation due to radius of curvature, coupling inefficiency, cladding and jacket design effects, and nonlinear effects due to severe thermal and mechanical environments.

1.6.1 Material Absorption Loss

Studies performed by various research scientists [6] indicate that material absorption loss is the most important one because almost all of this loss can be attributed to certain ions present in the glass fiber core. The studies further indicate that only a certain variety of glass compositions yield very small material absorption losses if no ion impurity is present in the composition. There are numerous metallic ions in a glass composition, such as Cr^{3+}, Cu^{2+}, Fe^{2+}, and OH^-, as illustrated in Figure 1–4. These ions have electronic transitions in the 0.5-to-1.0 micron spectral range. It is evident from Figure 1–4 that the absorption bands are of variable spectral width depending on the element involved. It is further evident that absorption loss due to the OH^- ion is the lowest, except at a wavelength of 0.85 micron. The peak absorption wavelength and width of the band could be different for a given ion in various glass compositions. Further, the valence state of a given ion influences the peak absorption wavelength and the absorption bandwidth. For example, iron with plus three state (Fe^{+3}), not shown in Figure 1–4, causes peak absorption below 0.4 micron. Thus, the details of the glass-making process used in the development of a glass fiber have profound effects on the absorption loss versus wavelength plot.

Another important absorbing ion in fused silica glass is the OH^- ion, which has the fundamental absorption peak near 2.7 microns (not shown in Figure 1–4). Note the peaks shown in Figure 1–4 at 0.95 microns and 0.72 microns are known as the second and third overtones of that vibrational absorption [6]. It is not necessary for the overtones to be exact harmonics of the fundamental, which are normally observed in the case of radio-frequency (RF) sources. However, the location of the peaks from the OH^- ion is strictly dependent on the ion concentration in a glass composition. Note other transition metallic ions will exhibit absorption losses at specific wavelengths depending on their ion concentrations. The above studies reveal that high purity in the glass is absolutely necessary for minimum absorption loss. The studies further reveal that unclad SM fibers made from bulk crystalline-quality glass exhibit significantly lower losses over a wide spectral bandwidth. Zinc chloride glass offers minimum insertion loss and bulk absorption coefficients over a wide spectral range, as illustrated in Figure 1–3.

1.6.2 Losses Due to Scattering Mechanisms

There are various types of scattering mechanisms that can cause insertion loss in an optical fiber. These mechanisms include Rayleigh scattering, Mie scattering, Stimulated Raman scattering, and Stimulated Brillouin scattering. The Rayleigh scattering phenomenon will always be present in a glass medium, and the coefficient of scattering is independent of the light-wave field strength.

1.6.2.1 Rayleigh Scattering

Rayleigh scattering is caused by thermal fluctuations, compositional variations, and phase separation. Rayleigh scattering energy appears both in the cladding and core. However, energy is absorbed in the core as a backward-scattered guided wave. As stated before, Rayleigh scattering is always present, and it is most dominant in an optical fiber below 1600 nm. The Rayleigh scattering coefficient is inversely proportional to the fourth power of the operating wavelength [7]. Rough calculations performed by the author indicate that the scattering loss is about 3 dB/km at 0.83 microns, 0.5 dB/km at 1.1 microns, 0.25 dB/km at 1.3 microns and 0.19 dB/km at 1.4 microns. Scattering loss in a fiber along with other losses as a function of wavelength is shown in Figure 1–6.

1.6.2.2 Mie Scattering

Mie scattering is caused by inhomogeneities comparable in size to operating wavelength and generally causes predominantly forward scattering. Loss in the fiber due to Mie scattering is significantly low regardless of wavelength.

1.6.2.3 Stimulated Raman and Brillouin Scattering Phenomena

Both the Stimulated Raman and Brillouin scattering are caused by nonlinear interaction between the traveling wave and the material properties, which is observed above a certain threshold power-density level. In the case of long-distance optical transmission, the nonlinear effects constitute an upper limit on the optical power that can be transmitted with minimum scattering. It is important to mention that Raman scattering is predominantly in the forward direction, whereas Brillouin scat-

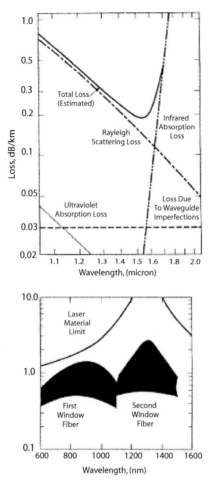

Figure 1-6 *Loss mechanisms in optical fiber and bandwidth capability as a function of operating wavelength.*

tering is predominately in the backward direction. In the case of Raman scattering in vitreous silica at a wavelength of 1 micron, the maximum usable light power is related to transmission loss in the fiber by Equation 1.3:

$$P_{max} = [(0.04)(IL)(w^2)] \text{ watts} \qquad 1.3$$

where IL is the insertion loss in the fiber (dB/km) and w is the full field width at half-maximum power density in the fiber (microns).

Note the field width (w) is used instead of the core size. In the case of an SM fiber where the nonlinear effects are more important, the electric field is not confined to the core. Note Equation 1.3 is valid either for a single frequency or a multifrequency optical source, provided the full-width parameter (w) is equal to or less than 0.01 micron or 10 nm [7]. In the case of a light-emitting diode (LED), the parameter w is less than 0.03 microns or 30 nm and the P_{max} will be about three times greater than the amount given by Equation 1.3. Computed values of maximum usable light power as a function of the parameters involved are summarized in Table 1–3.

As stated earlier, stimulated Brillouin scattering is predominantly in the backward direction. The critical power or the maximum optical power will vary depending on whether the light source is single-frequency or multifrequency. For a single-frequency source, the maximum usable optical power in an SM fiber can be written as

$$P_{max} = [(8 \times 10^{-5})(IL)(w^2)] \text{ watts} \qquad 1.4$$

Calculated values of maximum optical power are summarized in Table 1–4.

These computed values of maximum optical power indicate that very small power (mw) is scattered in the backward direction under stimulated Brillouin scattering phenomenon compared to large power (watts) scattered in the forward direction under stimulated Raman scattering phenomenon. The amount of power scattered in the forward or backward direction is strictly dependent on the type of optical source involved, for example a single-frequency or multifrequency light source.

Maximum usable optical power from a multifrequency source is dependent not only on the insertion loss (IL) and full width (w) parameters, but also on the source bandwidth ($\Delta\lambda$) which is expressed in angstrom (Å). The maximum usable optical power for a multifrequency source can be written as

$$P_{max} = [(0.08)(IL)(w^2)(\Delta\lambda)] \text{ watts} \qquad 1.5$$

Computed values of maximum power under the stimulated Brillouin scattering phenomenon for a multifrequency source as a function of insertion loss, full width, and source bandwidth are given in Table 1–5.

These computations indicate that the maximum usable power for a multifrequency source under stimulated Brillouin scattering phenomenon increases with an increase in all the variables. However, the increase is significant for a multifrequency source with higher spectral width or source bandwidth.

1.6 Various Loss Mechanisms in Optical Fiber 17

Table 1-3 *Computed values of* P_{max} *at a wavelength of 1 micron (watt).*

IL (dB/km)	Full width, w (micron)			
	1	2	3	4
1	0.04	0.16	0.36	0.64
2	0.08	0.32	0.72	1.28
3	0.12	0.48	1.08	1.92
4	0.16	0.64	1.44	2.56

Table 1-4 *Maximum optical power from single-frequency source in an SM optical fiber (mw).*

IL (dB/km)	Full Width, w (micron)			
	1	2	3	4
1	0.16	0.64	1.44	2.56
2	0.24	0.96	2.16	3.84
3	0.32	1.28	2.88	5.12
4	0.40	1.60	3.60	6.40

Table 1-5 *Maximum optical power for a multifrequency source under stimulated brillouin scattering phenomenon (watt).*

IL (dB/km)	Full Width, w (micron)							
	$\Delta\lambda=1$ (Å)				$\Delta\lambda=1$ (Å)			
	1	2	3	4	1	2	3	4
2	0.16	0.64	1.44	2.56	0.32	1.28	2.88	5.16
3	0.24	0.96	2.16	3.84	0.48	1.92	4.32	7.68
4	0.32	1.28	2.88	5.12	0.64	2.56	5.76	10.24
5	0.40	1.60	3.60	6.40	0.80	3.20	7.20	12.80

1.6.3 Scattering from Geometrical Variations

Geometrical variations in the size of an optical fiber core can lead to transfer of optical power due to scattering between guided modes and/or from a guided mode to the radiation field. This kind of scattering does not contribute significantly to the absorption effects or absorption loss in the core. Optical power distribution among the guided modes and the scattering between the guided modes to the radiation field will be discussed in the second chapter. Studies performed by the author indicate that it is extremely difficult to compute the radiation losses in multimode dielectric waveguides. The studies further indicate that the scattering power-loss coefficient for a specific mode is dependent on the refractive indices of core and cladding, correlation length, core diameter, and the propagation constant for that specific mode involved in the axial direction of the fiber. Dissipation loss that is not equal in all modes will prevent the steady-state distribution from occurring in the core. A systematic deviation of 0.1% in the core size will cause an exchange of energy between the lowest-order mode and the next mode that can occur in the fiber length not exceeding 5 cm. However, in a well-designed optical fiber, scattering effects of a higher magnitude do not occur. Scattering losses in an optical core or waveguide can be minimized, if not avoided, by maintaining a uniform fiber cross-section within a tolerance of less than 0.1% and by allowing the occurrence of residual core-size variations with very long periods.

1.6.4 Radiation Loss Due to Radius of Curvature

A dielectric waveguide or optical fiber will radiate if it is not absolutely straight. For a gradual bend or for a large radius of curvature (R), the transmission field in the radial plane will differ slightly from the normalized field associated with the straight section of the waveguide or optical fiber. The radiation attenuation coefficient (A_r) can be expressed as

$$A_r = [(C_1) \exp(-C_2 R)] \text{ dB/meter} \qquad 1.6$$

where C_1 and C_2 are constants and are independent of the bend radius or radius of curvature (R). The constant C_1 is of the order 10,000, whereas C_2 is of the order 100. Since the electric field in the cladding medium extends to a large distance, the introduction of a radius of curvature R implies that the energy is propagating at greater speed than the velocity of light. The radiation loss is dependent on the index difference parameter (Δ), which is defined by Equation 1.7. One can use this concept to calculate radiation loss associated with losing energy at the farthest location in the cladding.

$$\Delta = (n_1^2 - n_2^3)/\ 2n_1^2 \qquad 1.7$$

A typical value of the index difference is around 0.001. Index differential parameters greater than 0.001 will lead to smaller curvature losses. Studies performed by various research groups in the 1980s indicate that a bend radius on the order of 1 cm or better is possible.

1.6.5 Losses Due to Cladding Effects

Some losses can be expected due to Rayleigh scattering in the cladding material. Reconversion of the optical power from cladding modes into core-guided modes will have serious effects on the scattering losses and group delay. The scattering loss in the cladding will be considered large for cladding modes, but negligible for core modes. In addition, excessive loss in a jacket can also introduce some additional loss for the core-guided modes. If a lossless jacket is used and if the refractive index of the material outside the cladding is larger than the index of the cladding material (n_2), one can expect significantly lower loss for the core-guided modes. The cladding-design effects have significant impact on cross-talk between parallel fibers in a multimode optical cable.

1.6.5.1 Impact of Cladding Effects on Normalized Power Carried by Core and Cladding

Power-handling capability of cladding is seriously affected by the losses in cladding, which in turn can affect the power-handling capability of the core. The lowest-order mode carries the maximum power in both the core and cladding materials. The total power, which is the sum of core and cladding power components that is carried by an optical fiber, is defined as

$$P_{core} + P_{clad} = 1 \qquad 1.8$$

$$P_{core}/P + P_{clad}/P = 1 \qquad 1.9$$

where P_{core} and P_{clad} indicates the power carried by the core and cladding materials, respectively. It is evident from Equation 1.9 that the sum of normalized power carried by the core and cladding is unity. The normalized power components carried by the core and cladding can be computed by considering the power carried by all the modes that can be defined by Equation 1.8. These normalized power components [7] are dependent on the normalized propagation constant b, a constant u function of full wave parameter (w), and a normalized frequency parameter v. The normalized propagation constant can be written as

$$b = 1 - (u/v) \qquad 1.10$$

Studies performed by the author indicate that the maximum power density for the lowest-order mode occurs when the normalized frequency parameter v approaches 1.85 (approximately), whereas for the higher-order modes, the maximum power density occurs at higher values of v. The explanation for various values of the normalized frequency parameter will be provided in the second chapter. Computed values of normalized power in the core and cladding using assumed values of various constants for the lowest-order mode are summarized in Table 1–6.

Table 1–6 *Normalized power level carried by the core and cladding for the lowest-order mode*[*]*.

v	b	(u/v)	u	Normalized Power (%)	
				(P_{core}/P)	($P_{cladding}/P$)
0.5	0.00	1.000	0.500	100	00
1.0	0.08	0.949	0.960	62	38
1.5	0.21	0.894	1.342	31	69
2.0	0.32	0.830	1.662	11	89
2.5	0.51	0.712	1.782	7	93

[*] Note these are values are accurate within +/−10 because of various assumptions made.

1.7 Dispersion in Optical Fibers

A dispersion effect in optical fibers is considered very serious because it can seriously degrade the performance of long-haul wavelength division multiplexing (WDM) and dense-WDM (DWDM) transmission and communications systems [8]. Three distinct types of dispersion effects are critical: the material dispersion caused by index variations as a function of wavelength; cumulative dispersion due to bidirectional transmission of optical signals in the case of MM fibers; and chromatic dispersion due to strain, which is dependent on the operating temperature, source wavelength, coefficient of thermal expansion of the material, and surface conditions of the fiber.

1.7.1 Material Dispersion

Group-delay distortion sets a limit on the information rate that can be reliably transmitted in an optical transmission system. Material dispersion, which is caused by the variation of the refractive index as a function of wavelength, contributes to the group delay. Material dispersion in SM fibers fed with a narrow-band optical source is significantly low. However, material dispersion becomes a matter of serious concern for long-haul DWDM transmission systems. Dispersion poses a serious problem in the cases of multifrequency source and MM mode optical fibers. Typical dispersion in a standard SM fiber is close to 16 ps/nm-km at an operating wavelength of 1550 nm [8]. Under steady-state thermal and normal temperature pressure (NTP) environments, the chromatic dispersion can be called as a material dispersion. Material dispersion characteristics of core and cladding are very similar, and they can cause specific group delay (second/meter) in the optical signal. Modal dispersion associated with various modes will be discussed in the second chapter.

One can compute the material dispersion for an SM mode optical fiber made from fused silica glass using the index versus wavelength, index slope ($dn/d\lambda$), and second derivative of index slope ($d^2n/d\lambda^2$) curves depicted in Figure 1–7.

Material dispersion in an optical fiber can be expressed as

$$D_m = (\lambda/c)(d^2n/d\lambda^2) \qquad 1.11$$

where λ is the wavelength, c is the velocity of light, and n is the refractive index of the core material. This index for a fused silica glass fiber is 1.451 at 1000 nm wavelength and 1.446 at 1550 nm. As stated earlier, the refractive index for the core material varies as a function of wavelength. It is necessary to find the second derivative of the index of a specific material with respect to wavelength to determine the value of a material dispersion for a specific fiber-core material. Using the second-derivative values shown in Figure 1–7, one can compute the material dispersion in a SM optical fiber with a fused silica core. Computed values of material dispersion in an SM fiber with fused silica core are summarized in Table 1–7.

Studies performed by the author on various SM fibers reveal that a standard step-index SM fiber has dispersion close to zero at 1310 nm, whereas its attenuation typically is between 0.3 to 0.4 dB/km. However, a standard SM fiber has material dispersion close to 16 ps/nm-km at 1550 nm, whereas its insertion loss typically varies from 0.2 to 0.3 dB/km [7]. The material dispersion value of 16 ps/nm-km can be verified both from Figure 1–8 and Table 1–7.

Optical fibers with controlled dispersion are available for additional cost. High dispersion poses a serious problem, particularly for high-speed data-transmission optical lines. High dispersion not only affects high-speed data transmission capability, but also the optical signal density. These problems can be

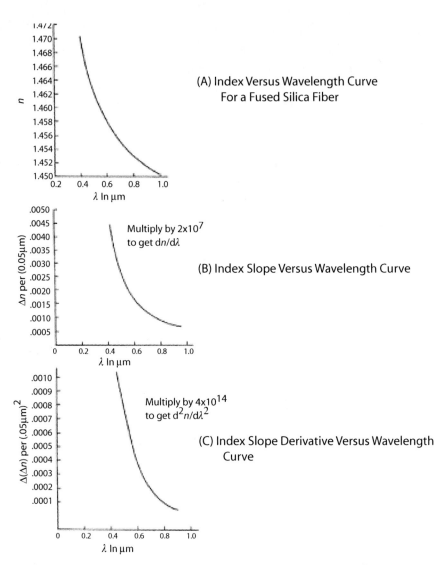

Figure 1–7 *Various curves used to estimate the dispersion in SM fibers.*

overcome by controlling the dispersion, its slope, and the area of the fiber. Dispersion control boosts the high-speed data-transmission capability of an optical system. Reduced material dispersion is possible by altering the refractive-index profile of the fiber. A zero-dispersion-shifted fiber (see Figure 1–8) offers zero dispersion at 1550 nm, but it has limited application due to the presence of nonlinear effects, which makes it unsuitable for a WDM telecommunications

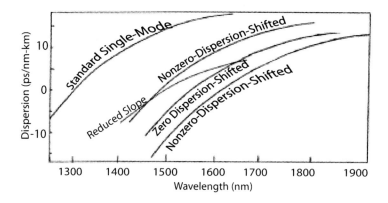

Figure 1–8 *Dispersion curves for SM fibers using various dispersion compensation techniques as a function of operating wavelength.*

Table 1–7 *Material dispersion computations for SM fibers (ps/nm-km) [9].*

Wavelength, λ (nm)	$(d^2n/d^2\lambda)$ X (4×10^{14})	Material Dispersion (ps/nm-km)
800	0.0001000	21.2
1000	0.0000185	18.4
1200	0.0000853	17.2
1400	0.0000831	16.6
1600	0.0000822	16.4

system. It is important to mention that too little dispersion spreads the signals and reduces their interactions. On the other hand, too much dispersion limits the transmission distance and may require dispersion-control schemes. Optimum performance requires a fiber with a balanced amount of material dispersion, namely, 5 to 10 ps/nm-km. Projected material-dispersion curves as a function of wavelength for SM fibers are shown in Figure 1–8.

1.7.2 Chromatic Dispersion

Chromatic dispersion in an optical fiber is strictly dependent on optical wavelength, operating temperature, CTE of the core material, and surface conditions of the core. Its typical value varies from 0.1 to 5 ps/nm-km in the 1530-to-1560 nm spectral regions and under steady-state thermal environments. Studies performed in the early 1980s reveal that complication in optical systems comes from the slope of the dispersion curves (dn/dλ) shown in Figure 1–7. The typical dispersion slope for an SM fiber is about 0.08 and 0.09 ps/nm^2-km at 1550 nm and 1310 nm operation, respectively. Higher dispersion slopes are not considered suitable for long-haul transmission systems.

1.7.3 Cumulative Dispersion

Cumulative dispersion only occurs in MM optical fibers. This type of dispersion is due to bidirectional transmission of optical signals and is dependent on the number of modes traveling in each direction and the optical properties of the MM fiber elements. Cumulative dispersion does not cause signals to spread in time rather than in wavelength. However, pulse spreading caused by the cumulative dispersion does not limit the maximum data that can be transmitted over a length of fiber.

1.7.4 Potential Dispersion-Compensation Schemes

Dispersion-compensation schemes are critical in applications where high data transmission over a wide spectral region and over a large distance is the principal requirement. In other words, dispersion compensation is absolutely essential in long-haul DWDM telecommunications systems. Note the total dispersion in the case of SM fibers includes the chromatic dispersion and the material dispersion components. Controlling or compensation of total dispersion in a fiber requires an exotic compensation scheme. The latest scientific research studies [10] indicate that a transmitter and a receiver could get away with no compensation for a 100 km transmission using state-of-the-art SM fibers. However, they may require some compensation for 300 km spans and elaborate compensation for spans exceeding 800 km. The studies further indicate that Bragg gratings are best suited for chromatic dispersion compensation. The studies further conclude that lower-dispersion wavelengths (close to 1550 nm) can go five times further than higher-dispersion wavelengths (close to 1310 nm) before any dispersion compensation is required.

For a zero-dispersion-shifted fiber, the dispersion can be written as

$$D(\lambda) = [(S_0)(\lambda_0^4 / \lambda^3)] \qquad 1.12$$

where S_0 is the zero-dispersion slope and λ_0 the zero-dispersion wavelength. Assuming a zero-dispersion slope of 0.09 ps/nm^2-km, zero-dispersion wavelength of 1310 nm, and zero-dispersion-shifted wavelength of 1550 nm,

Equation 1.12 yields a dispersion value of 0.179 ps/nm-km for an SM, zero-dispersion-shifted optical fiber. This illustrates a unique dispersion compensation scheme for SM fibers. Most SM fibers used in communications networks are DS fibers (Figure 1-8) specially designed for use at 1550 nm (C-band operation). However, by making the zero-dispersion wavelength 1550 nm (i.e., non-zero-dispersion-shifted wavelength), undesirable nonlinear effects such as self-phase modulation and stimulated Brillouin scattering are generated. One can eliminate such nonlinear effects by moving to a higher wavelength of 1580 nm.

1.8 Polarization Loss

Polarization loss occurs in a fiber when the amplitude of the electric field is not equal for both the horizontal and vertical polarization. One will never find a material that will provide a perfect polarization-preservation (PP) capability. Therefore, there always will be some polarization loss, no matter how small. Polyimide or acrylate coating on the optical fiber provides a polarization-maintaining capability or a PP capability. The polyamide and acrylate coatings offer low polarization loss under bending or crushing or kinking. Corning "PURE-MODE-PM" engineered fiber (PM-1300/400 or PM-1500/400) provides minimum attenuation, extremely high birefringence, low sensitivity to bending, high resistance to crushing and kinking, and best PP capability. The coating film outer diameter varies from 250 to 400 microns for optical fibers operating over the 1290-to-1450 nm spectral range. PP single-mode (PPSM) optical fibers are available with optimum performance over the 820-to-870 nm and 1300-to-1550 nm spectral regions. PPSM fibers with a beat length as small as 2 mm are available at moderate cost. The beat length is defined as the length of the fiber over which polarization rotates through 360 degrees. Studies performed by the author [8] indicate that the shorter the beat length, the better the PP properties of the fiber irrespective of the operating wavelength. The studies further indicate that a "Bow-Tie" cladding architecture offers high PP capability with stress-induced intrinsic birefringence and polarization loss less than 0.05. SM fibers with a high-extinction ratio and PP capability are readily available.

1.9 Summary

Material requirements for the core, cladding, buffer, and inner and outer jackets are defined. Mechanical, electrical, and optical properties of various fiber elements are summarized. Performance capabilities and limitations of SM and MM optical fibers are briefly discussed. Potential loss mechanisms are identified. Fiber losses due to absorption, scattering, polarization, and coupling inefficiency are described as a function of wavelength. Mathematical expressions to compute material dispersion, cumulative dispersion, and chromatic dispersion are derived. Stimulated Raman scattering and stimulated Brillouin scattering dispersion for

SM and MM fibers are discussed. Maximum power-handling capability as a function of insertion loss, full width and spectral bandwidth are computed for a single-frequency source and multifrequency source. Radiation loss due to the radius of curvature is discussed. Dispersion for a zero-dispersion-shifted SM fiber is computed as a function of zero-dispersion slope and zero-dispersion wavelength at a specific optical wavelength.

1.10 References

1. Jha, A. R. (1993, July). *Oxynitride glass fibers* (technical proposal). Cerritos, CA: Jha Technical Consulting Services.
2. Jha, A. R. (2000). *Infrared technology: Applications to electro-optics, photonic devices, and sensors* (p. 169). New York: John Wiley and Sons, Inc.
3. Contributing Editor. (2000, July). Space-qualified 1.2 mm fiber optic cable. *Microwave Journal*, 156–158.
4. Jha, A. R. (1991, September). *High temperature substrate materials for SMART-SKIN technology* (technical report). Cerritos, CA: Jha Technical Consulting Services.
5. Miller, S. E., et al. (1973). Research towards optical-fiber transmission systems. *The Proceedings of IEEE, 61*(12), 1704–1750.
6. Miller, S. E., et al. (1973). Research towards optical-fiber transmission systems. *The Proceedings of IEEE, 61*(12), 1708.
7. Jha, A. R. (2001, August). *Fiber optic cables and connectors* (technical proposal). Cerritos, CA: Jha Technical Consulting Services.
8. Jha, A. R. *Infrared technology: Applications to electro-optics, photonic devices, and sensors* (p. 176). New York: John Wiley and Sons, Inc.
9. Jha, A. R. (1994, July). *Programmable microwave fiber-optic delay line network* (technical report). Cerritos, CA: Jha Technical Consulting Services.
10. Hecht, J. (2000, July). Dispersion control boosts high-speed transmission. *Laser Focus World*, 107–109.

CHAPTER 2

Expressions for Electric Field, Propagating Modes, Group Delay, and Coupling Coefficients

E xpressions for electric fields, propagating modes, group delays, and coupling coefficients will be derived because of their critical importance to understanding the performance capabilities and limitations of optical fibers. Potential scalar modes and hybrid modes in an optical fiber will be discussed. The Transverse Electric (TE) and Transverse Magnetic (TM) modes are not of practical importance and are much easier to describe than the hybrid modes. Mathematical expressions for optical power carried by the core, cladding and jacket will be derived for the most prominent hybrid mode. Power-level computation supporting the hybrid modes such as HE_{11} or EH_{11} needs the complex form of the dyadic function for a rigorous formulation and analysis. In the case of optical communication fibers, the magnitude of the electric field, radiation loss, and power levels in various modes are strictly dependent on surface irregularities, frequency or wavelength of surface roughness, variations in optical fiber radius, and dielectric differences between the core and cladding. High computation accuracy for the electric field and modal power levels is only possible when the optical fibers have a very small dielectric difference between the core and cladding. Radiation loss due to random variation in the core radius along the fiber length is critical, particularly in long-haul communication fibers. Power loss due to trapped-surface modes will be briefly investigated.

Expressions for the electric field and modal power will be developed at cutoff for the TE and TM modes. Modal power produced by the core surface irregularities generates forward and backward scattering modes, leading to development of the hybrid modes. An asymptotic expression for the radiation power involving Bessel functions (J_0 and J_1) and Hankel functions (K_0 and K_1) will be derived as a function of wavelength of surface roughness, unperturbed core radius, and normalized optical frequency. Conditions for maximum power transfer between two modes will be specified. Attenuation of radiation fields in higher-order modes will be investigated.

The exact number of propagating modes in a circular fiber is complicated and cumbersome and requires numerical solution for the eigenvalues. However, very simple and highly accurate asymptotic expressions for the eigenfunctions and eigenvalues, which are valid for all frequencies when the dielectric difference between the core and cladding is small (less than 0.01), need to be derived. Since most formulations for excitation, radiation, and scattering are based on approximations identical to those used in deriving the asymptotic expression, it is not necessary to manipulate the exact equations involved. Application of the asymptotic expressions to scattering and excitation is presented for indicating a multimode propagation phenomenon along the optical fiber axis or a circular dielectric cylinder. Eigenvalue equations are complicated transcendental equations involving Bessel and Hankel functions. Conditions for normalization or scaling procedures are specified. Discussion of the zero-order eigenvalue equation is necessary because it provides an accurate asymptotic zero-order solution for all frequencies. Boundary conditions for rapid decay of the electric field or maximum power propagation within the core are summarized.

Both the symmetrical and hybrid modes are excited in an optical fiber depending on the core and cladding radii, core surface condition, and angle of incident. At small incidence angles, several hybrid modes are excited. Even a small imperfection in a dielectric waveguide or optical core has a significant impact on the dominant hybrid mode HE_{11}. Modal power levels as a function of fiber aperture, incident angle, and dielectric constant differential are computed for transverse and longitudinal modes. Conditions are specified for propagation for the hybrid mode HE_{11}, thereby restricting scattering to this hybrid mode, in addition to a radiation field. The power scattering by an incident HE_{11} mode back into the same hybrid mode HE_{11} is discussed in greater detail as a function of critical fiber parameters. Computed values of radiated- and scattered power levels are provided as a function of normalized frequency parameter V.

2.1 Electromagnetic (EM) Fields and Propagating Modes

The asymptotic expressions for the EM fields for the geometry of a uniform circular dielectric cylinder or round optical fiber (see Figure 2–1) can be defined by the eigenfunction and the eigenvalue equations. Due to the geometry of a uniform circular dielectric guiding structur,e shown in Figure 2–1, the asymptotic expression for the EM field can be defined by the eigenvalue and eigenfunction equations using a normalizing process.

Since the propagation constant is a function of the circular core radius (ρ), angular frequency (ω), and fiber geometrical parameter (δ), it is convenient to define the normalized frequency (V) by 2.1.

$$V^2 = [U^2 + V^2] or [u_p^2 + v_p^2] \qquad 2.1$$

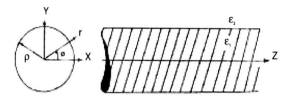

Symbols:
ρ = Core Radius, r = Distance From the Center of the Optical Fiber, Ø = Angle of the Ray From X-axis
ε_1 = Dielectric Constant of the Core and ε_2 = Dielectric Constant of the Cladding

Figure 2–1 *Uniform circular optical fiber structure showing the core and cladding regions.*

The longitudinal EM field components are of paramount importance and can be defined by the normalized wave equation.

The solution of the normalized wave equation for the pth can be given as

$$[\rho^2 \nabla^2 + (u_p^2 \text{ or } -w_p^2)] = 0$$
$$\text{use } u_p^2 \text{ when } R \leq 1 \text{ and } -w_p^2 \text{ when } R \geq 1 \qquad 2.2$$

$$\nabla_t = [\nabla^2 - \partial^2/\partial z^2]$$
$$\nabla = \text{del operator} = [i(\partial/\partial x) + j(\partial/\partial y) + k(\partial/\partial z)] \qquad 2.3$$

where i, j, and k are unit vectors in the direction of the x, y, and z coordinates, respectively, R is the normalized radius equal to (r/r), and r is the distance from the center, as illustrated in Figure 2–1.

Due to the cylindrical symmetry of the optical fiber (see Figure 2–1), the EM vector fields for the pth can be expressed as

$$\mathbf{E}_{pl}(x,y,z) = [\mathbf{e}_p(x,y) + \mathbf{E}_{pz}(x,y)]e^{jx}$$
$$\mathbf{H}_{pl}(x,y,z) = [\mathbf{h}_p(x,y) + \mathbf{H}_{pz}(x,y)]e^{jx} \qquad 2.4$$

where $x = [\text{wt} + \beta_{pz}/\rho]$, $\mathbf{e}_p$ is the transverse electric-field vector, $\mathbf{h}_p$ is the transverse magnetic-field vector, $\mathbf{E}_{pz}$ is the longitudinal electric field vector, and $\mathbf{H}_{pz}$ is the longitudinal magnetic-field vector for the pth mode. The parameter β_{pz} is the normalized modal constant for the pth mode, which is defined as

$$\beta_p^2 = [k^2(1-\theta_p^2)] \qquad 2.5$$

$$\theta_p = [(\sqrt{\delta})(U_p/V_p)] = [u_p/k] \qquad 2.6$$

$$k = [2\pi/\lambda] \qquad 2.7$$

The parameter k is the normalized propagation constant for a z-directed plane wave in the core medium with a dielectric constant of ε_1, λ is the wavelength in the same medium, and the parameter u_p is the eigenvalue of the normalized wave 2.2 for the pth mode. The eigenvalues defined by this equation are best suited for an asymptotic analysis according to Jones [1]. The EM fields for the hybrid modes (HE_{lm} or EH_{lm}) can be defined as a function of longitudinal or azimuthal variation (l) and radial variation (m) by appropriate equations. When the azimuthal variation is zero, the electric and magnetic field degenerate to the TM_{0M} and TE_{0M}, which are defined below.

For TE_{0M} modes, the derivatives of the longitudinal electric and magnetic fields along the z-axis can be defined as

$$E_z' = [J_l(uR)]g_1(\phi) \quad \text{when } R < 1$$

$$= [\eta_3 K_l(wR)g_1(\phi)] \quad \text{when } R > 1 \qquad 2.8$$

$$H_z' = [(\sqrt{\varepsilon_1\mu})(F_2\beta/k)][g_1(\phi)/g_2(\phi)]E_z'$$

$$g_1(\phi) = [\sin l\phi \text{ or } \cos l\phi]$$

$$g_2(\phi) = [\cos l\phi \text{ or } -\sin l\phi]$$

$$\eta_3 = [J_l(u)/K_l(w)]$$

$$\phi = \text{Azimuthal angle as shown in Figure 2.1.}$$

$$F_2 = (V/uw)^2[1/\eta_1 + \eta_2]$$

$$\eta_1 = [J_l'(u)/uJ_l(u)]$$

$$\eta_2 = [K_l'(w)/wK_l(w)] \qquad 2.9$$

The prime notations on the Bessel function $J_1(u)$ and modified Hankel function $K_1(w)$ indicate the derivatives with respect to the arguments u and w, respectively. It is important to mention that when the azimuthal variation (l) is equal to zero, the electric and magnetic fields degenerate to the TH_{om} and TE_{om} set modes. For the TE_{om}, the derivatives of the longitudinal electric- and magnetic-field expressions along the z-axis can be written as

$$E_z^{'} = [J_0(uR)] \text{ for } R < 1$$
$$= [\eta_3 K_0(wR)] \text{ for } R > 1 \qquad 2.10$$

$$H_z^{'} = [(\beta/k)(\sqrt{\varepsilon_1/\mu}) J_0(uR)] \text{ for } R < 1$$
$$= [(\beta/k)(\sqrt{\varepsilon_1/\mu}) \eta_3 K_0(wR)] \text{ for } R > 1 \qquad 2.11$$

where R is the normalized radius equal to (r/ρ), $J_0(uR)$ is the Bessel function of zero order, $K_0(wR)$ is the modified Hankel function of zero order, μ is the permeability of the medium, ε_1 is the dielectric constant of the core material, and k is equal to ($2\pi/\lambda$).

2.2 Impact of Normalization Procedure

Using the normalizing procedures involving the U and W parameters, one obtains the eigenvalue equation for all modes, which can be written as

$$[UJ_l(U)/J_{l\mp1}(U)] = \pm[WK_l(W)/K_{l\mp1}(W)] \qquad 2.12$$

The upper sign (+) indicates the HE_{lM} hybrid mode, whereas the lower sign (−) indicates the EH_{lM} hybrid mode. When the azimuthal variation is equal to +1, the eigenvalue equation of HE_{1M} can be written as

$$[UJ_1(U)/J_0 U)] = [WK_1(W)/K_0(W)] \qquad 2.13$$

When the azimuthal variation is equal to −1, the eigenvalue equation for the EH_{1M} hybrid mode can be written as

$$[UJ_1(U)/J_0(U)] = -[WK_1(W)K_0(W) \qquad 2.14$$

2.3 Impact of Cutoff Conditions on Eigenvalue Equations

The normalized frequency parameter V for a pth mode can be defined as

$$V = [\sqrt{U_p^2 + W_p^2}] \qquad 2.15$$

For a circular optical fiber, the simplest and approximate expressions for the EM fields associated with an HE_{11} hybrid mode can be used to derive the analytical expressions for fractional power components and using dimensionless, normalized frequency parameter V. This normalized frequency parameter can be written as

$$V = [(2\pi\rho_1/\lambda)\sqrt{n_1^2 - n_2^2}] \qquad 2.16$$

where ρ_1 is the core radius (Figure 2–1), λ is the wavelength of the light in a vacuum (0.606 micron), n_1 is the refractive index of the core material, and n_2 is the refractive index of the cladding material. Assuming a core made from fused silica with a refractive index of 1.450, a core diameter of 3.0 micron, and cladding made from pure silica with a refractive index of 1.442, one gets the value of the normalized frequency as follows:

$$\begin{aligned}V &= [(2\pi \times 1.5/0.606)\sqrt{1.4502^2 - 1.4418^2}]\\ &= [15.5587 \times \sqrt{2.1031 - 2.0788}]\\ &= [15.5587 \times 0.1559]\\ &= [2.4254]\end{aligned}$$

This value is very close to 2.405 for the hybrid mode HE_{11} for a fused silica core. High accuracy in computation of V is only possible when the differential index or permittivity parameter (δ) approaches much less than 0.01. This parameter is defined as

$$\delta = [1 - (\varepsilon_1/\varepsilon_2)] = [1 - (n_2^2/n_1^2)] \qquad 2.17$$

Where ε indicates the permittivity or dielectric constant, n indicates the refractive index of the medium, and subscripts 1 and 2 represent the core and cladding, respectively. It is important to mention that the dielectric constant of the core and cladding materials change slightly with the optical frequency, whereas radical change in the refractive index is observed at higher wavelengths as illustrated in Table 2–1 [2].

Table 2–1 *Refractive-index change for fused silica core as a function of wavelength.*

Wavelength (λ), micron	Refractive Index (n_1)
0.5	1.4619
1.0	1.4502
1.5	1.4447
2.0	1.4372
2.5	1.4309
3.0	1.4234

As stated earlier, the cutoff conditions define the values of the eigenvalue equation, the modal power within the dielectric core, and the normalized group velocity. The near-cutoff condition for the normalized frequency (V) is defined by the condition when W_p approaches zero, whereas the far-cutoff condition for the normalized frequency (V) is defined by the condition when W_p is much greater than 1 ($V \gg U_p$).

For a zero-order eigenvalue equation, parameter u_p is equal to U_p, and U_p becomes highly accurate when the quantity θ_p^2 given by Equation 2.6 is neglected. From the definition of θ_p as defined by Equation 2.6, it is confirmed that far from cutoff θ_p^2 it is much less than 1 for all values of parameter δ, but at cutoff it is equal to δ. Therefore, θ_p, defined as the characteristic angle that a cone of a plane wave makes with the dielectric boundaries to form the pth mode, is the largest at cutoff if parameter $\delta \ll 1$. Under this condition, the asymptotic zero-order solution U_p is valid at all optical frequencies, i.e., for all values of parameter V. In fact, the values of parameter U_p obtained using Equation 2.12 have an error less than 1% at all frequencies when parameter δ is equal to 0.2 and less than 10% when δ is equal to 0.5. This occurs when the core permittivity is twice that of the cladding material [3]. Numerical solution for the eigenvalue equation U_p requires asymptotic forms of Hankel functions K_1 and the derivative expression for the above parameter. Having said this, one can now write the derivative for the parameter U_p as

$$U_p' = [dU_p / dV] = (U/V)[1 - 1/\xi] \qquad 2.18$$

The parameter ξ is frequently used when a circular dielectric rod study is involved. It can be written as

$$\xi = [K_l K_{l\mp 2}]/K_{l\mp 1}^2 \qquad 2.19$$

Using the asymptotic format and modified Hankel functions, 2.19 can be rewritten as

$$\xi = [1+1/V] \text{ when } V >> U_p \qquad 2.20$$

Substituting 2.20 into 2.18 results in

$$[dU_p/dV] = [U_p/V^2] \qquad 2.21$$

which has a solution of

$$U_p(V) = [U_p(\infty)e^{-1/V}] \qquad 2.22$$

where $U_p(\infty)$ are the roots of $J_{l+/-1}$. This means [4]

$$\begin{aligned}
U_p(\infty) &= \text{roots of } J_{l\mp 1} \\
&= 2.405\, HE_{11} \\
&= 3.932\, TM_{01}, TE_{01}, HE_{21} \\
&= 5.135\, EH_{11}, HE_{31} \\
&= 5.520\, HE_{12} \\
&= 6.370\, EH_{21}, HE_{41} \\
&= 7.016\, TM_{02}, TE_{02}, HE_{22}
\end{aligned}$$

Values of the normalized eigenvalue (U_p) can be calculated using either 2.12 or 2.22. Computed values for various modes shown in Figure 2–2 seem to be in excellent agreement except near cutoff. Since optical guides or transmission lines are operated above cutoff [4], the parameter U_p given by 2.22 is an excellent representation when the parameter θ_p^2 is small. As stated previously, the parameter θ_p is very small far from cutoff and equal to $\sqrt{\delta}$ at cutoff. Therefore, when parameter $\delta << 1$, the propagating modes at any frequency are formed by nearly z-directed plane waves, as illustrated in Figure 2–1.

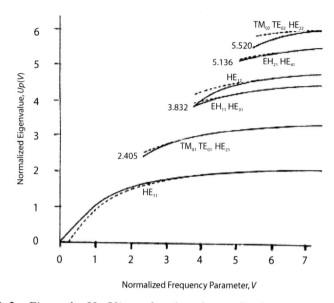

Figure 2–2 *Eigenvalue* $U_p(V)$ *as a function of normalized frequency for various modes present in an optical fiber. Note lines offer exact solutions, whereas dashed lines yield approximated solutions for the pth mode.*

2.4 Linearly Polarized Modes

A simplified characteristic equation for linearly polarized (LP) modes was developed by Gloge [5]. The LP modes are known as pseudomodes because they do not yield an exact solution. The number of propagating modes increases as the size of the fiber increases relative to wavelength. Note various group modes have different field configurations but have very nearly the same propagation constant (β). The characteristic equation is valid on the assumption that the cladding diameter is so large that the outer region of refractive index of n_0 has no impact on the wave propagation.

The characteristic equation involving a quantity b can be written as

$$[uJ_{l-1}(u)/J_l(u)] = (-w)[K_{l-1}(w)/K_l(w)] \qquad 2.23$$

$$b = [1-(u/v)^2] \qquad 2.24$$

2.24 defines the quantity called normalized propagation constant in terms of normalized parameters, namely, u and v. Values of the normalized propagation constant

(b) for three distinct modes in an optical fiber as a function of the normalized frequency parameter (V) are shown in Figure 2–3. This normalized propagation constant has a zero value at cutoff and approaches unity as the wavelength (λ) approaches zero.

2.5 Physical Significance of Parameter θ_p

As stated earlier, this particular parameter is the characteristic angle that a cone of plane waves makes with the dielectric boundaries when the normalized radius R is less than 1 to form the pth mode. The value of this parameter is very small far from cutoff frequency but equal to the square root of the parameter δ. When the dielectric difference between the optical core and the surrounding layer (δ) is much less than 1, the modes at any optical frequency are formed by the z-directed plane wave. Since the parameter θ_p has the largest value at cutoff when $\delta<1$, the asymptotic zero-order solution U_p is valid for all optical frequencies, i.e., for all values of parameter V. As stated earlier, values of the U_p parameter computed by 2.12 will have an error of less than 1% when parameter δ is less than 0.2 and less than 10% when δ is equal to 0.5.

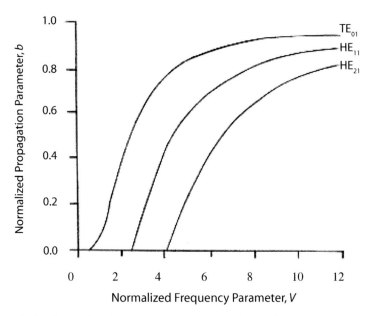

Figure 2–3 *Normalized propagation parameter* b *as a function of normalized frequency parameter* V *for specific dominant modes in the optical fiber.*

2.6 Computations of Normalized Eigenvalues (U_p) as a Function of Normalized Frequency (V)

Normalized eigenvalues are strictly dependent on the dominant operating mode, normalized frequency, and the roots of the Bessel function $J(U)_{l+/-1}$ (given by 2.22 for various modes). As mentioned previously, θ_p is very small and equal to the square root of parameter δ at cutoff ($\delta = 1-\varepsilon_2/\varepsilon_1$). Therefore, the computed normalized eigenvalues using 2.22 will be most accurate for all frequencies V if parameter $\delta \ll 1$.

Table 2–2 *Computed normalized eigenvalues (U_p) for the HE_{11} hybrid mode.*

Normalized Frequency (V)	Normalized Eigenvalue U_p	Normalized Parameter W_p
2.405	1.586	1.807
4.000	1.873	3.534
5.000	1.972	4.596
6.000	2.036	5.644
7.000	2.085	6.682
8.000	2.122	7.713
9.000	2.152	8.739
10.000	2.176	9.760

2.7 Impact of Parameter δ on the Accuracy of Eigenvalues

The differential dielectric or refractive-index parameter δ, as defined by 2.17, determines the computational error in the normalized eigenvalues. Computed errors as a function of dielectric constant or permittivity (ε) and the refractive index (n) for the core indicated by the subscript 1 and for the cladding by the subscript 2 are shown in Table 2–3.

These computations assume fused silica for the core and cladding materials. The computations indicate that the errors in normalized eigenvalues are less than 1% if the δ is less than 0.02 and less than 10% when its value approaches 0.5.

Table 2–3 *Impact of parameter δ on errors associated with eigenvalues.*

ε_2	ε_1	δ	Error (%)	n_2	n_1	δ	Error (%)
3.68	3.78	0.026	2.2	1.440	1.455	0.021	0.62
3.12	3.77	0.178	3.5	1.436	1.451	0.019	0.54
2.65	3.76	0.297	6.8	1.432	1.444	0.016	0.38
1.92	3.75	0.500	9.6	1.425	1.435	0.014	0.29
$\delta = [1-(\varepsilon_2/\varepsilon_1)]$				$\delta = [1-(n_2/n_1)^2]$			

2.8 Number of Propagating Modes in Optical Fiber

The number of propagating modes are strictly dependent on the numerical aperture, concentric or eccentric configuration of the core with respect to cladding, incident angle, core-cladding boundary, core radius, index profile configuration and the refractive indices of the core and cladding materials, and small-index differential parameter (δ). If there are N free-space modes in operation, there will be the same number of propagating modes in the circular fiber. Selected circularly symmetrical modes, such as TE_{ol} and TM_{ol}, are less affected by the eccentricity than the higher-order noncircular symmetrical hybrid modes, such as HE_{lm} and EH_{lm}. Note the HE_{11} mode belongs to circularly symmetric modes and is the most dominant hybrid mode in a circular optical fiber.

The number of modes in a circular, bend-free optical fiber can be written as

$$N = [2(\theta/\delta)^2]$$
where $\delta = [\lambda/\pi a]$ 	2.25

and λ is the wavelength in vacuum and a is the core radius.

However, wave theory yields nearly the same number of modes involving other parameters and can be given as

$$N = [(kan_1)^2 \Delta]$$
where $k = [2\pi/\lambda]$
and $\Delta = [(n_1 - n_2)/n_1]$ 	2.26

2.8 Number of Propagating Modes in Optical Fiber

The maximum acceptance angle (θ) of the fiber is defines as

$$sin\theta = [\sqrt{n_1^2 - n_2^2}] \approx [n_1\sqrt{2\Delta}] \qquad 2.27$$

$$N_{p.index} = [(kan_1)^2](\Delta/2) \qquad 2.28$$

Equation 2.28 yields the number of modes within a fiber with a parabolic-index profile. Computed values of propagating modes using Equations 2.25, 2.26, and 2.28 are summarized in Table 2–4 using the given parameters.

The refractive indices of 1.440 and 1.435 have been assumed for the core and cladding materials, respectively. It is interesting to mention that both Equations 2.25 and 2.26 yield approximately the same number of modes.

The number of modes within a parabolic-index fiber is equal to half of the modes propagated in a single clad fiber as shown in the last column. Therefore, an optical fiber must use an optimized parabolic-index value if a minimum number of propagation modes is the principal requirement.

The total number of modes as a function of the normalized frequency (V) range [6] is summarized in Table 2–5. These modes account for two polarizations of each symmetric and nonsymmetric mode types wherever applicable.

Table 2–4 *Propagating modes in a sinple clad optical fiber.*

a (micron)	θ	δ	$(kan_1)^2$	Δ	Number of Modes		
					EQ 2.25	EQ 2.26	EQ 2.28
2	.2069	.0964	904	0.0132	9.32	9.30	4.66
3	.2069	.0642	2034	0.0132	20.97	20.95	10.49
4	.2069	.0482	3616	0.0132	37.25	37.25	18.65
5	.2069	.0386	5650	0.0132	58.28	58.21	29.15

Table 2–5 *Total number of propagating modes as a function of normalized frequency (V).*

Normalized Frequency Range	Additional Modes	Total Number of Modes
0–2.405	HE_{11}	2
2.405–3.832	TE_{01}, TM_{01}, HE_{21}	6
3.832–5.136	HE_{12}, EH_{11}, HE_{31}	12
5.136–5.521	EH_{21}, HE_{41}	16
5.521–6.380	TE_{02}, TM_{02}, HE_{22}	20
6.380–7.016	EH_{31}, HE_{51}	24
7.016–7.588	HE_{13}, EH_{12}, HE_{32}	30
7.588–8.417	EH_{41}, HE_{61}	36

2.8.1 Impact of Fiber Bending on Propagating Modes

The number of propagating modes or guided modes in a multimode fiber is affected by the bending of its axis. The number of guided modes for a step-index fiber and parabolic-index profile fiber [6] can be written as

$$N(R) = [\{1-(a/R\Delta)\}] \qquad 2.29$$

where R is the bending radius, a is the core radius, Δ is a parameter defined earlier by 2.26, N is the number of modes without bend, and $N(R)$ is the number of modes with a bend in the fiber.

2.8.2 Coupling Between Two Guided Modes

The periodic coupling between two guided modes can be defined by

$$C(z) = [A\sin(\theta z)] \qquad 2.30$$

where $C(z)$ is the coupling coefficient, A is the amplitude, z is the distance along the z-axis, and θ is a propagation constant difference between the pth mode and q^{th} mode, which can be written as

$$\theta = [\beta_p - \beta_q] \qquad 2.31$$

Complete transfer of power from one mode to the other occurs when the difference between propagation constants is equal to the coupling parameter defined by 2.31. Note this coupling parameter between two mode pairs will be extremely sensitive. In MM optical fibers, a mode coupling will have a spectral response defined as

$$C(z) = [\sum_{n=1}^{\infty} A_n sin(\theta_n z)] \qquad 2.32$$

where A_n is the coupling amplitude and θ_n is the coupling parameter for the nth mode.

Studies performed by the author on optical fibers indicate that the core diameter could have a spectrum variation as a function of parameter z. For small coupling magnitude or perturbation, the round-fiber theory and the slab-guide theory yield comparable results. The studies indicate that a round or circular optical fiber will have radiation loss much larger than the slab-guide radiation loss. On the other hand, coupling is strictly dependent on the correlation length. In the absence of loss, the short-correlation-length coupling produces a steady-state distribution of the optical power in the lowest-order mode. Coupling with a long correlation length yields a steady-state distribution of power in all modes equally. However, the dissipation loss that is not equal in all modes will prevent these power distributions from occurring. A systematic deviation of the core wall, even of 0.1% of the core diameter, will result in a total exchange of energy between the lowest-order mode and the next mode in a fiber length not exceeding 5 cm. Coupling loss as a function of the coupling parameter and other relevant parameters is depicted in Figure 2–4 for two different polarization states.

2.9 Crosstalk Between Optical Fibers

Crosstalk between the coupled fibers is critical when a MM fiber is used in long-distance optical communications systems. The higher the crosstalk, the poorer the quality of voice communication will be. Therefore, reduction of crosstalk becomes a matter of paramount importance in long-distance communication. Crosstalk is dependent on the coupling coefficient between the fibers, jacket material, and separation between fibers and jacket thickness, as in the expression $(R-2b)$, where R is the separation between the two fibers and b is the jacket radius. Preliminary calculations indicate that significant reduction in crosstalk

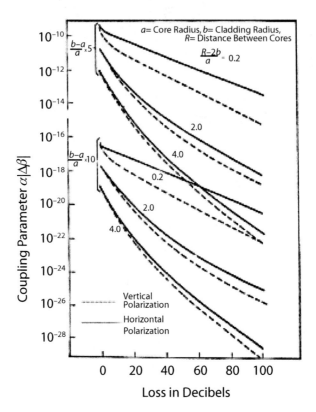

Figure 2–4 *Coupling parameter between two coupled fibers versus a loss experienced by a plane wave going from one fiber boundary to another fiber boundary.*

can be achieved using a lossy jacket. The lossy jacket is characterized by the parameter L, which is given as

$$L = [(b-a)/a] \qquad 2.33$$

where b is the cladding radius and a is the core radius.

Reduction in crosstalk (i.e., normalized loss) as a function of parameter L for the dominant hybrid mode HE_{11} in an optical fiber is evident from the curves shown in Figure 2–5. The higher the parameter L, the lower the crosstalk will be. A reduction by an order of seven is possible by increasing the L from 5 to 10. Undesired loss for the core-guided modes is introduced by the jacket. The lossy

2.9 Crosstalk Between Optical Fibers

jacket occupies the gap between the two fibers and has a thickness equal to $R-2b$ along the x-axis, as shown in Figure 2–5, where R is the distance between the two fibers and b is the cladding radius. Cladding thickness of five times the core radius yields the quantity $2\alpha a$ close to 8×10^{-7} for the abscissa (Figure 2–5), which is equivalent to about 1 dB. For a core radius of 2 micron, the α parameter has a value close to 0.25 dB/meter or 250 dB/km besides the HE_{11} hybrid mode loss due to the lossy jacket. Assuming cladding thickness of ten times the core radius instead of five times, the HE_{11} is smaller by a factor of 10^6, as illustrated in Figure 2–5.

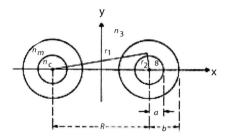

(a) Cross-section of Two Coupled-Cladded Fibers Separated by Distance R and Using a Jacket Thickness of $(R-2b)$, where b is the Cladding Radius

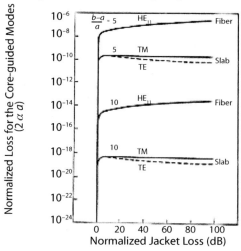

(b) Lossy Jacket Used to Reduce Crosstalk Between Two Fibers

Figure 2–5 *(a) Cross-section of two coupled fibers and (b) reduction of crosstalk.*

2.9.1 Impact of Propagation and Core Radius on Crosstalk

The magnitude of crosstalk can be defined by the crosstalk ratio (V), which is expressed as

$$V = [D/\Delta\beta]^2 \qquad 2.34$$

where D is the distance at which the unit power incident in one fiber yields a power of V in the other fiber and $\Delta\beta$ is the change in propagation constant due to the crosstalk between two degenerated modes. The quantity $a\Delta\beta$ is called the coupling parameter, also known as the crosstalk ratio, and is plotted against insertion loss [7] in Figure 2–4 for various polarization states and core/cladding dimensions. These curves are generated for $n_m ka = 16$, $\gamma a = 1.594$ and $n_c/n_m = 1.01$. From these curves one can conclude that if a crosstalk ratio (V) of 60 dB can be tolerated (i.e., $V = 10^{-6}$), and if a core radius of 2 micron and a distance of 1 km are assumed, 2.34 yields the quantity $\Delta\beta$ not exceeding 10^{-9} per meter and the product $a\Delta\beta$ less than 2×10^{-15}. Figure 2–4 illustrates that to achieve the $a\Delta\beta$ of 2×10^{-15} with a cladding thickness of five times the core radius requires at least 15 dB loss as shown by abscissa. However, by doubling the cladding thickness, i.e., ten times the core radius, an additional drop of 70 dB in crosstalk is achieved with no lossy jacket. It is easy to prevent higher core-guided mode loss through a lossy jacket of appropriate thickness. A lossy jacket is not very effective in reducing direct crosstalk between the core-guided modes; a cladding medium of appropriate thickness is a better choice.

2.10 Group Delay in Optical Fibers

Optical energy propagation in the various modes is affected by the presence of group delay. In other words, group delay has a serious impact on the overall performance of optical fibers.

The group delay (T_{gd}) associated with the wave guidance [8] can be expressed as

$$T_{gd} = (L/c)[d(n_1 k)/dk + (n_2 \Delta)d(vb)/dv] \qquad 2.35$$

where L is the optical fiber length, c is the velocity of light in vacuum, v is the normalized frequency, and b is the normalized propagation constant at cutoff which approaches to unity as λ_o approaches to zero.

The propagation constant (b) is defined as

$$b = [1 - (u/v)^2] \qquad 2.36$$

$$v = [(ka)\sqrt{n_1^2 - n_2^2}] \qquad 2.37$$

Note all other parameters have been defined earlier. The first term in 2.35 is known as delay dispersion due to the bulk material from which the fibers are made. It is the same for all modes present in the core. The second term is called group delay and is associated with the wave guidance. Its derivative can be written as

$$[d(vb)/dv] = [1-(u/v)^2(1-2K')] \qquad 2.38$$

$$K' = [K_l^2(w)/K_{l-1}(w)K_{l+1}(w)] \qquad 2.39$$

$$v = [\sqrt{u^2 + w^2}] \qquad 2.40$$

The normalized group delay is defined by 2.38. Note $K(w)$ is the Hankel function with "w" argument, while K' is a parameter defined by Hankel functions mentioned in 2.39. Since, HE_{11} is the most important dominant hybrid mode in an optical fiber, group delay values using 2.38 and 2.40 have been computed for this particular mode These are shown in Table 2–6 Using the above equations, one can obtain values of normalized group delay for other modes.

Computed values of normalized group delay as a function of normalized frequency (V) for three distinct modes are shown in Figure 2–6.

Table 2–6 *Computed values of normalized group delay for HE_{11} hybrid mode.*

v	b	(u/v)	u	w	d(vb)/dv
2.5	0.032	0.984	2.462	0.434	-------
3.0	0.181	0.906	2.719	1.268	1.092
4.0	0.422	0.761	3.044	2.598	1.352
5.0	0.611	0.624	3.122	3.905	1.259
6.0	0.721	0.529	3.174	5.092	1.202
7.0	0.772	0.477	3.339	6.152	1.162
8.8	0.815	0.431	3.442	7.223	1.144

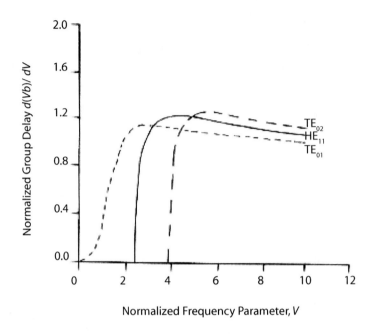

Figure 2–6 *Normalized group delay as a function of parameter V for various modes.*

Once the values of the normalized propagation constant (b) are obtained using the curves, as explained in the article by Gloge [8], values of other parameters such as u, v, and w can be computed for the circular symmetrical modes and noncircular hybrid modes. Computed values of the normalized propagation constant (b) for three distinct dominant modes, as a function of normalized frequency (V), are shown in Figure 2–3.

2.11 Group Delay Spread

Group delay spread in an MM optical fiber is critical. The spread or variation of group delay for the modes present in an MM fiber is dependent on the refractive-index differential parameter Δ. Small group delay requires small values of refractive-index differential parameter ($\Delta \ll 1$) and high values of normalized frequency the ($v \gg 1$). The group delay spread dT can be written as

$$dT = [(n_1)(\Delta)(L)/c][1 - 2/v] \qquad 2.41$$

$$\text{where } \Delta = [(n_1 - n_2)/n_1] \qquad 2.42$$

2.11 Group Delay Spread   47

Assuming the refractive-indices values of 1.45 for the core and 1.44 for the cladding, fiber length (L) of 5 cm, normalized frequency (v) of 2.405 for the dominant hybrid mode HE_{11} mode, the differential-index parameter (Δ) of 0.0069 and velocity light c of 3×10^{10} cm/sec, one expresses the group delay spread as

$$dT = [1.45 \times 0.0069 \times 5/2 \times 1010][1-2/2.405]$$

$$= [0.056] \text{ ps/cm or } [5.6] \text{ ps/meter.}$$

The above equations indicate that the refractive index of the core and the parameter Δ have significant impact on the group delay spread in a fiber of specific length.

Computed values of group delay spread as a function of refractive indices of core and cladding for an optical fiber 5 cm long are shown in Table 2–7.

These computations indicate that a slight increase in the cladding index can make a significant impact on delay spread. When the cladding index increases to 1.435 from 1.420, the group delay spread is doubled. This indicates that tight control of the cladding refractive index is critical if minimum delay spread is desired. In summary, for minimum group delay spread in a MM fiber, the refractive-index differential parameter Δ must be kept well below 0.01 or 1%.

Table 2–7 *Computed values of group delay spread as a function of refractive indices of the core and cladding layers.*

n_1	n_2	Δ	Spread (ps/meter)
1.450	1.440	0.0069	5.62
1.450	1.435	0.0103	8.36
1.450	1.430	0.0138	11.21
1.450	1.420	0.0207	16.82

2.12 Modal Power Within Optical Fiber

The modal power levels for various modes in an optical fiber are well described by Alan Snyder [2]. The modal power within a fiber or dielectric circular rod can be expresses by parameter η which is defined as

$$\eta = (U/V)^2 \{[(W/U)^2 + 1/\xi] \qquad 2.43$$

$$\xi = [K_l K_{l\mp2}/(K_{l\mp1})^2] \qquad 2.44$$

Near cutoff (when $W \ll 1$), the parameter ξ is equal to $2 \log_e (2/1.78W)$ when $l=0$ and $l=2$ for HE_{lM} mode. However, for far above cutoff ($W \gg 1$ and $V \gg U$), its value is simply equal to $1+ 1/V$. This means that 2.43 at far above cutoff is reduced to

$$\eta = [1 - (U/V)^2] \qquad 2.45$$

for the hybrid mode HE_{lM}.

Modal power levels in the fiber can be calculated as a function of normalized U and V parameters. These computations reveal that maximum modal power levels are present in the dominant hybrid mode HE_{11}, as shown in Figure 2–7. The lowest modal power levels will be found in the HE_{31} and EH_{11} hybrid modes. This indicates that modal-power calculations must be performed with greater accuracy for the most dominant hybrid mode HE_{11}.

2.12.1 Modal Power Carried in the Core and Cladding Layers

The distribution of modal power carried by the optical fiber components is very important. If the total modal power carried in a specific mode is equal to P, assuming no power carried by the jacket, the expressions for power carried in the core and cladding layers can be written as

$$[P_{core}/P] = [1 - (u/v)^2(1-\kappa)] \qquad 2.46$$

$$[P_{clad}/P] = [(u/v)^2(1-\kappa)] \qquad 2.47$$

$$\kappa = [K_l^2(w)/K_{l-1}(w)K_{l+1}(w)] \qquad 2.48$$

$$v = [\sqrt{u^2 + w^2}] \qquad 2.49$$

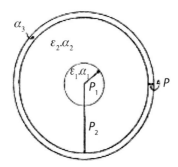

(A) Ideal Circular Optical Fiber Showing the Core Radius, Cladding Radius, Cladding Thickness and Loss Constants

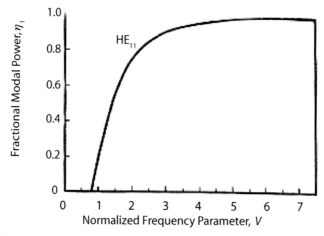

(B) Fractional Modal Power in HE_{11} Mode within the Core as a Function of V

Figure 2–7 *(a) Ideal optical fiber configuration and (b) modal power in HE_{11} mode.*

Once the parameters u, v, and κ are known, the normalized modal power in the core and cladding layers can be calculated using the equations above. Values of normalized propagation parameter b, as a function of v, can be taken from the curves shown in Figure 2–3, values of parameter u from the curves shown in Figure 2–2, values of parameter w from 2.49, and values of parameter κ as a function of w from 2.48 (involving modified Hankel functions). Computed values of normalized modal power in core and cladding layers as a function of parameters u, v, b, and κ are shown in Table 2–8.

Table 2–8 Normalized modal power carried by the core and cladding layers for the dominant hybrid mode HE_{11}.

v	b	u	w	κ	P_{core}/P (η_1)	P_{clad}/P (η_2)
2.4048	0.52	1.662	1.736	0.655	0.819	0.181
3.8317	0.75	1.926	3.312	0.763	0.928	0.072
5.1356	0.85	1.976	4.736	0.785	0.968	0.032
5.5201	0.87	1.994	5.149	0.808	0.975	0.025

2.12.2 Power Density as a Function of Various Parameters

The power density expression for the dominant mode in the core [5] can be written as

$$P_d(a) = [(\kappa)(u/v)^2(P/\pi a^2)] \qquad 2.50$$

where parameter a is the core radius and P is the total modal power. For the lowest-order modes, the maximum power density occurs when v is equal to approximately 1.8. However, for the higher-order modes, the maximum power density in the core occurs at large values of parameter v.

The power density in the cladding layer at a distance r can be expressed as

$$P_d(r) = [(\kappa)(u/v)^2(P/\pi ar)][\exp(-2w(r-a))/a] \qquad 2.51$$

where r is much greater than a and other parameters are defined previously. Note for all modes except the lowest azimuthal-order mode, power in the cladding decreases with the distance from the core, even at cutoff conditions [5].

2.13 Fractional Power in Core, Cladding, and Jacket Layers

In the case of a long-distance optical communication system, the HE_{11} surface mode or hybrid mode, which is the dominant mode in a circular dielectric core, must propagate with minimum loss. In other words, the losses in the cladding and jacket layers must be kept to a minimum.

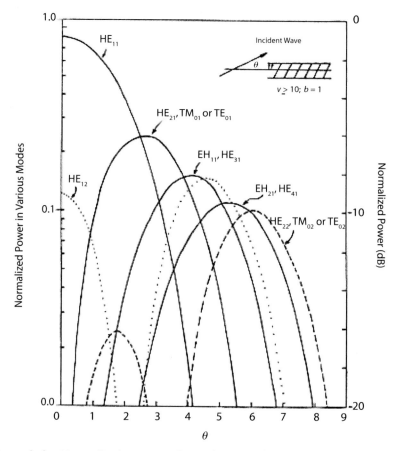

Figure 2-8 *Normalized power in the modes excited by a uniform source for small dielectric difference between the core and cladding and for small θ.*

The sum of fractional power levels in the core, cladding, and jacket is equal to unity. Therefore, one can write an expression for a circular, cladded optical fiber with a lossy jacket as

$$[\eta_1 + \eta_2 + \eta_3] = 1 \qquad 2.52$$

where η_1, η_2, and η_3 are the fractional modal power levels in the core, cladding, and jacket, respectively. Note as the cladding radius or thickness approaches infinity, the fractional modal-power component in the jacket is reduced to zero.

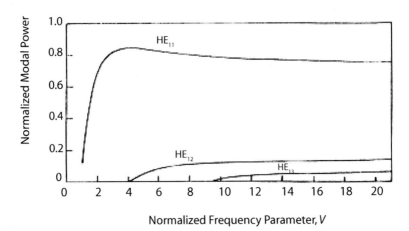

Figure 2–9 *Modal power due to excitation of modes at normal incidence by uniform fields on the circular end only. All the hybrid modes are excited; however, the first three hybrid modes are depicted under unity input power to the circular fiber core.*

Thus, one can simply write the expression for the fractional modal power level in the cladding as

$$\eta_2 = [1 - \eta_1] \quad 2.53$$

Fractional modal power levels for the cladding and core using 2.53 are shown in Table 2–8. These values assume a zero modal power level in the jacket.

2.13.1 Fractional Modal Power Due to Lossy Jacket

The fractional modal power within the lossy jacket can be given as

$$\eta_3 = [(0.5)(U/V)^2][\pi/W^2 K_1^2(W)][(1-e^{-x})(e^{-y})]$$
where $x = [2W(\Delta P/P_1)]$
$y = [W(P_2/P_1)]$ \quad 2.54

Sample Calculation for η_3:
  Assuming $\Delta\rho/\rho_1 = 0.1$, $\rho_2/\rho_1 = 10$, $U = 2.405$
  $\exp(-1/V) = 2.405 \times 0.6598 = 1.5868$,
  $W = \sqrt{U^2 - V^2} = \sqrt{5.7840 - 2.5179} = 1.8072$, $K_1(1.8072) = 0.1802$, and
  $(U/V)^2 = (1.5868/2.405)^2 = 0.4353$, one gets,

$$\eta_3 = [0.5 \times 0.4353][\pi/(3.266 \times 0.0325)][(1-0.6967)(1.417 \times 10^{-8}]$$
$$= [0.2176][29.609][0.3033 \times 1.417 \times 10^{-8}]$$
$$= [6.44 \times 0.423 \times 10^{-8}]$$
$$= [2.768 \times 10^{-8}]$$

From Table 2–8, $\eta_1 = 0.819$, $\eta_2 = 0.181$, and $\eta_3 = 2.768 \times 10^{-8}$ (by calculation), and their sum comes to unity, which satisfies Equation 2.50.

2.13.2 Attenuation Constant for the Dominant Hybrid Mode HE_{11}

The overall attenuation constant γ for the HE_{11} mode can be written as

$$\gamma = [0.819\alpha_1 + 0.181\alpha_2 + 2.767 \times 10^{-8}\alpha_3] \qquad 2.55$$

where, α_1 is bulk loss in core, α_2 is bulk loss in cladding, and α_3 is bulk loss in the jacket.

2.14 Summary

Complex mathematical expressions for electric and magnetic fields, propagating modes, group delay, group delay spread, and coupling coefficients for various optical fibers are derived. Dominant scalar and hybrid modes in an optical fiber are discussed. Fractional optical power levels carried in the core, cladding, and jacket have been computed for the hybrid modes using complex forms of the dyadic function, capable of yielding rigorous formulation and analysis. High computational accuracy is achieved when the optical fibers with very small dielectric difference between the core and cladding are selected for the computations. Normalized eigenvalues for the dominant hybrid mode HE_{11} are computed as a function of normalized frequency parameter V. The number of propagating modes as a function of core radius, core and cladding refractive indices, and wavelength are estimated using two different formulas. Coupling between two guided modes and crosstalk between two adjacent optical fibers are computed as a function of core, cladding, and jacket dimensions. Group delay and its spread are calculated as a function of the normalized parameters. Modal power levels carried in the core and cladding layers are computed using normalized parameters, Bessel and modified Hankel functions. The attenuation constant for the dominant mode HE_{11} in a round fiber has been calculated using fractional modal power levels and bulk losses in the core, cladding, and jacket.

2.15 References

1. Jones, A. L. (1965). Coupling of optical fibers and scattering in the fibers. *American Journal of Optical Society, 56*, 261–267.
2. Snyder, A. W. (1969). Asymptotic expressions for eigenfunctions and eigenvalues of a dielectric or optical waveguide. *MTT, 17*(12), 1130–1144.
3. Jones, D. S. (1964). *The theory of electromagnetism* (pp. 167–183). London: Pergaman Press.
4. Snyder, A. W. (1969). Asymptotic expressions for eigenfunctions and eigenvalues of a dielectric or optical waveguide, *MTT, 17*(12), 1130–1144.
5. Miller, S. E., et al. (1973). Research towards optical-fiber transmission systems. *The Proceedings of IEEE, 61*(12), 1703–1742.
6. Gloge, D. (1972). Bending loss in multimode fibers with graded and ungraded cores. *Applied Optics, 11*, 2506–2513. .
7. Marcuse, D. (1971). The coupling of degenerated modes in two parallel dielectric waveguides. *Bell System Technical Journal, 50*, 1791–1816.
8. Gloge, D. (1971). Weakly guiding fibers. *Applied Optics, 10*, 2252–2258.

CHAPTER 3

Fiber Optic Passive Components

Fiber optic (FO)-based passive components have potential applications in optical long-distance communication, scientific research, photonic sensors, medical equipment, industrial systems, space sensors, and military weapons systems. Frequently used FO-based components and devices include directional couplers, attenuators, fiber lasers, variable-ratio polarization-maintaining couplers, spectral filters, isolators, circulators, combiners and splitters, polarization-dependent loss compensators, polarization scramblers, limiters, and high-speed optical switches. High-speed switches include microelectromechanical systems (MEMS), thermo-optical switches, FO-based cross-matrix switches, and photon switches. It is important to mention that certain FO-based components such as couplers, isolators, Bragg grating filters and cross-matrix switches are widely used in WDM and demultiplexing systems, as illustrated in Figure 3–1. The pass-band optical filter is the most critical element and is widely used in a variety of applications including electro-optic, optoelectronic, photonics, and infrared (IR) sensors. These filters have pass-band requirements ranging from a few angstroms to a few hundred angstroms. Performance requirements, capabilities, limitations, and reliability aspects of critical FO-based components and devices will be discussed. The latest application of FO technology called Fiber to the Home (FTTH) will be briefly reviewed. The FTTH technology has potential for new Internet/TV/phone appliances on top of the home "Triplexing" boxes that are capable of receiving and splitting the triple-play signals at each home. It takes a video signal seven hours to download by a cable modem, whereas FO-based FTTH technology requires *only* three minutes.

3.1 Optical Limiters

Optical limiters are used to protect the sensitive optical amplifiers and receivers from unexpected and sudden high-power optical pulses. The limiters only allow the transmission of low-intensity optical signals, while blocking the high-intensity signals. These devices use microlens arrays, porous optical substrates, and optical fibers of specific lengths. The new generation of optical-limiter designs will be based on sol-gel glass fabrication technology capable of

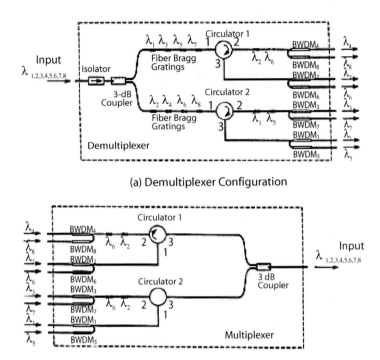

Figure 3–1 *Demultiplexing (a) and multiplexing system (b) configurations showing various FO components.*

providing very low threshold levels. Limiters with low threshold levels will be best suited for protection of delicate high-resolution optical sensors used by the military and space systems. In addition, optical limiters have potential applications in telecommunications, aerospace, and scientific research.

Each limiter must be optimized for the operating wavelength, IR spectral bandwidth, and threshold level. Optimum performance generally can be achieved over a narrow spectral bandwidth. However, both cost and complexity will increase with the increase in the spectral bandwidth.

3.2 Optical Filters

Optical filters are used to optimize the optical-signal parameters over a specified spectral bandwidth. Gain flattening filters are often used in the optical amplifiers, where excessive gain fluctuations are not acceptable. Fixed-frequency filters, diplexing filters, and tunable filters will be discussed, emphasizing performance, cost, and complexity. Fixed filters include low-pass filters, high-pass filters, and passband filters. Tunable filters are widely used in tunable receivers or tunable optical sources. Studies performed by the author indicate that tunable filters reduce the number of optical detectors from three to one, thereby realizing significant reductions in cost with no compromise in systems reliability. Performance requirements of filters for applications in erbium-doped fiber amplifiers (EDFAs), optical parametric oscillators (OPOs), WDM communications systems, and dense-WDM telecommunications systems will be briefly discussed with emphasis on insertion loss, skirt selectivity, and off-band rejection.

3.2.1 Low-pass (LP) Filters and High-pass (HP) Filters

These filters are widely deployed in applications where signals at specified wavelengths are allowed to pass with minimum attenuation, while signals at all other wavelengths are attenuated. In some applications, LP and HP filters can provide optimum performance with minimum cost and complexity. However, such applications are very limited. Optical filters with passband characteristics are best suited for multiple applications, including commercial, industrial, military, and space applications.

3.2.2 Band-pass (BP) Filters

BP filters are generally used for noise or spurious rejection over a specified spectral region, for gain-flattening functions, and for separation of the spectral band over a wide IR spectral region. BP filters, sometimes with stringent requirements, are used to meet critical performance requirements in demanding applications such as underwater, airborne, or space-based sensors. BP filters with low insertion loss, minimum dispersion, and sharp skirt selectivity are widely used in WDM and dense-WDM telecommunications systems to reduce cross-channel noise or crosstalk. BP filters are available for use in WDM or dense-WDM systems with 16, 32, 64, or higher channel capability. An SM BP filter operating at a wavelength of 1550 nm and with a 1 nm passband region has a typical insertion loss of about 1 dB, 3-dB bandwidth of 65 nm, back reflection of 055 dB, polarization-dependent loss (PDL) of 0.025 dB, and wavelength stability of 0.005 nm per degree Celsius. Significant performance improvement is possible at cryogenic temperatures.

3.2.3 Add-drop Filters Using On-Chip Technology

A two-dimensional digital micrometer (TDDM) can act as a fault-tolerant add-drop filter with potential applications in WDM and dense-WDM communications systems. The TDDM can be coupled by free-space optics to optical-fiber-connected array-waveguide-grating (AWG) multiplexers that form a two-dimensional in/out feed device. Currently most add-drop filters used in telecommunications systems rely solely upon on-chip technology integrated with waveguides and switches. This technology offers products with very high space bandwidth at minimum cost. A TDDM is comprised of more than one million micromirrors and involves macropixels involving several hundreds of micromirrors, each having switching capability at individual wavelengths. This design architecture is fault tolerant because failure of individual mirrors will have insignificant impact on the optical-alignment sensitivity due to overall robustness of the device. Add-drop filters must be developed at 1550 nm, the most popular wavelength, because of their great demand for application in WDM and dense-WDM telecommunications systems.

3.2.4 Heat-Absorbing (HA) Filters

Precise control of unwanted heat energy is critical in certain applications, such as entertainment lighting, medicine, optical communications, and ultraviolet curing. Excessive heat can degrade the performance of electro-optical sensors. Thin-film coatings of specific optical materials can separate the light from excessive thermal energy, thereby minimizing the heat-related performance degradation. Optical filters using heat-absorbing coatings are called heat-absorbing (HA) filters. These filters offer improved transmission of visible wavelengths, while rejecting thermal energy at IR wavelengths. When using these filters, it is extremely important to remove the heat energy absorbed by circulating cold air in the vicinity of the filters. Air cooling may not be sufficient for intense focused energy, and thus other cooling methods must be deployed to avoid catastrophic failure of the optical systems.

3.2.5 Tunable Optical Filters

Tunable optical filters are widely deployed in military imaging sensors, telecommunications systems and other scientific research applications. Several types of tunable filters are available in the market, but FO tunable filters (FOTFs) are receiving greater attention. These filters offer good performance with minimum cost and are widely used in WDM and dense-WDM telecommunications systems [1]. FOTFs are available with an instantaneous optical bandwidth ranging from 1 to 3 nm over the tuning range from 1290 to 1320 nm and a tuning range exceeding 30 nm at a spectral wavelength of either 1310 or 1550 nm. These filters have a maximum insertion loss of 1.25 dB, tuning resolution of 0.05 nm, back-reflection better than -55 dB, PDL of less than 0.03 dB at 1560 nm, and wavelength stability of 0.005 nm/°C over a temperature range of 0 to +50°C.

3.2.5.1 Applications of Tunable Filters in Commercial and Military Applications

A tunable filter offers maximum operational flexibility. It can select individual wavelength channels from a stream. Conventional demultiplexing schemes require several fixed-wavelength filters, leading to a costly and complex filter package. FOTFs offer an effective technique for increasing the information-carrying capacity of an optical WDM or dense-WDM communications system. FOTFs will be available within the next five years at minimum cost. FOTFs have potential applications in EDFAs for effective side-band suppression, tunable lasers for scientific research, wavelength-selecting technique for optical tests, and high-power lasers to focus maximum optical energy at a specific wavelength.

3.2.5.2 Various Tunable Filter Concepts

Tunable filters other than FOTFs are available for various applications, but they are, heavy and require large input power. A tunable notch filter is widely used in high-power lasers because it offers maximum eye and sensor protection from nearby-operating high-power tunable leasers. A tunable notch filter will block out unwanted laser radiation, while allowing the system to operate at a specific wavelength. These filters provide adequate protection to commercial and fighter aircraft pilots from the high-power lasers intended to blind them. Performance capabilities, including the tuning speed and the tunable bandwidth of the filter, determine its cost.

Liquid-crystal tunable filters are best suited for image detection and display applications because of low cost and minimum power consumption. Fabry-Perot tunable filters can be used in WDM and dense-WDM systems, when and if they are available at affordable costs. Acousto-optic (AO) tunable filters using piezo-electric tuning elements demonstrate remarkable performance capabilities for microscopy and remote sensing applications, but they suffer from excessive weight, high cost, and very high power consumption. State-of-the-art AO filters [2] demonstrate transmission efficiency in excess of 90%, a tuning period of less than one second, and a tuning resolution of much less than 0.5 nm over the 400-to-500nm spectral range. These filters are best suited for underwater detection systems, optical-imaging sensors, and astronomical systems.

3.2.6 Linear Variable Filters (LVFs)

Linear variable filter (LVF) technology is best suited for a variety of spectrometer applications. A spectroscope integrated with an LVF is designed for in-situ process control of gases, liquids, and solids. Such spectroscopes allow continuous control and monitoring of critical process parameters required for higher yields and improved quality control.

Pyroelectric IR detector arrays offer a crude form of LVF technology. A pyroelectric detector array with 64 pixels offers signal output, uncooled operation, and wavelength sensitivity from 1 to 20 microns. Each pixel is capable of detecting a different wavelength in various IR spectral ranges, such as 1.4 to 2.4 microns or 3 to 5 microns or 5.5 to 11 microns or 8 to 10 microns. However, its performance is very marginal, and such arrays are not suited for commercial and military applications, where tuning accuracy, resolution, and consistency are of paramount importance.

3.3 Optical Couplers

FO-based optical couplers are readily available in various configurations; such as 1×2, 2×2, 1×4, 1×8, or 1×16 and are capable of operating over narrow as well as wide spectral regions. Single-mode FO (SMFO) couplers are bidirectional devices that can be used either to split or combine optical signals with minimum loss. Single-wavelength-coupler designs are optimized to provide optimum performance at a specific and demanding operating wavelength of 633, 780, 850, 1310, or 1550 nm. Couplers operating either at the 1310 or 1550 nm wavelength are available with a bandwidth of +/-20 nm or lower. SMFO couplers with 1×2 and 2×2 configurations and coupling values of 3 dB are available at minimum cost because of straightforward design procedures. The FO T-coupler is equivalent of a microwave coupler. The fabrication of such couplers is extremely simple. The unclad fibers are joined by transparent cement with a refractive index closely matching that of the optical fiber. Most efficient coupling results when the refractive index of the cement is equal to that of the fiber. Additional coupling losses can occur because of discontinuities within the optical coupler formed by the mechanical misalignments. The coupling losses are dependent on the numerical aperture (NA) of the fiber, operating wavelength, and spectral bandwidth. It is evident from Figure 3-2 that the coupling losses decrease with the NA of the fiber.

3.3.1 FO Couplers for Military System Applications

The SMFO coupler configuration shown in Figure 3-3 is best suited for programmable delay lines for possible applications in electronic countermeasures (ECM) systems. Programmable delay lines are used to confuse the enemy's tracking radars and hostile homing missiles. The FO coupler architecture shown in Figure 3-3 has a typical loss of less than 0.5 dB and directivity greater than 55 dB. The PDL is less than 0.1% in SM fibers. SMFO couplers with minimum insertion loss, lowest POL, and maximum directivity over a desired spectral bandwidth must be designed to meet critical military performance requirements in harsh environments.

3.3.2 Optical Couplers for Telecommunications Applications

FO couplers are widely used in WDM and dense-WDM telecommunications systems where PDL and insertion loss fluctuations are critical. FO couplers with ultralow PDL, maximum insertion loss well below 0.5 dB, and isolation greater than 60 dB are best suited for WDM, dense-WDM, and polarization-sensitive

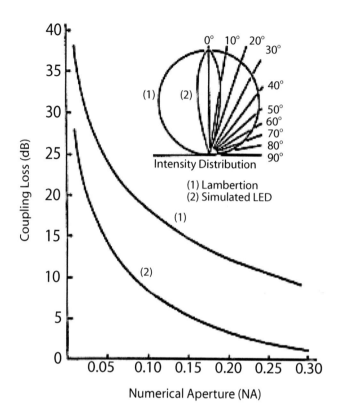

Figure 3–2 *Coupling loss for SM optical coupler as a function of NA parameter for two different optical sources.*

telecommunications systems. Specific internal details of a WDM coupler capable of accepting both the 1310 and 1550 nm optical signals containing voice and video data are shown in Figure 3.4.

3.4 Optical Isolators and Circulators

Optical isolators and circulators are used to improve the overall performance of optical systems in the presence of reflected signals. Optical reflections can degrade the performance of lasers, optical amplifiers, and other sensitive optical sensors if the reflected energy is not adequately blocked. Isolators and circulators provide adequate protection from the reflected optical signals.

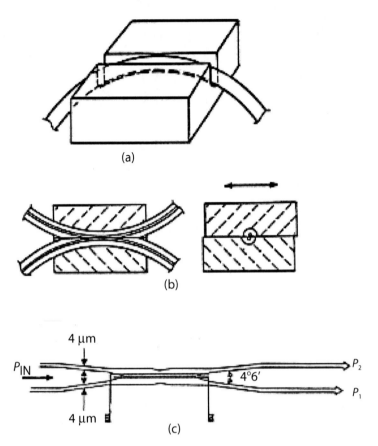

Figure 3–3 *(a) SM fiber directional coupler, (b) mated halves sections showing side-view (left) and end-view (right), and (c) architecture of a coherent optical coupler.*

3.4.1 Optical Isolators

An isolator is a passive, unidirectional optical device that allows the optical beam to transmit in the forward direction with minimum insertion loss, while blocking the signal transmission in the reverse direction with an attenuation ranging from 50 to 70 dB depending on the type of isolator deployed. Optical isolators of different design configurations are available, including all-fiber isolators, fiber-embedded isolators, fiber Faraday rotator isolators, and waveguide-based isolators for high-power applications. Commercially available Faraday isolators use a pair of birefringent crystals or a pair of polarizers. Microlenses or gradient-index lenses are used to couple the optical beam from one fiber to the other. The birefringent crystal-based

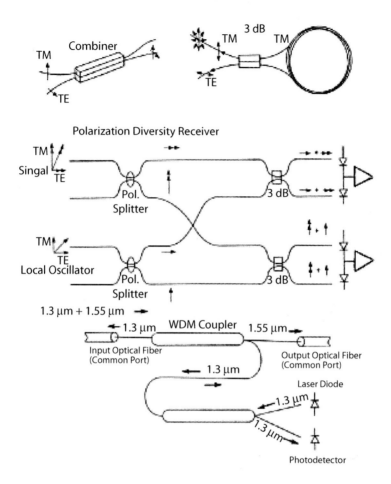

Figure 3–4 *Critical optical components of a communication system.*

structure is called a polarization-independent isolator, whereas the polarization-based isolator design is called a polarization-dependent isolator.

Most semiconductor diode lasers emit linearly polarized beams with extinction ratios close to 20 dB, which may not be adequate for some applications. Furthermore, SM fibers do not maintain the input polarization during the transmission. Polarization-independent isolators are mostly used in applications where a minimum isolation of 45 dB and maximum insertion loss of 0.05 dB are the principal requirements. In a polarization-independent isolator, the first wedged crystal splits the input laser beam into two parallel beams that are then refocused by a lens into an optical fiber in the forward direction. The optical beams, after passing through a 45-degree Faraday rotator, enter the second wedge crystal. The two beams exit the surface of the second wedge crystal parallel to each other. The optical lens focuses both beams into the output fiber. In the reverse direction, the second wedge again splits the incoming beam into two beams. After passing through the 45-degree Faraday rotator, both beams are rotated by 45 degrees in the same direction and the beam directions are interchanged in the first wedge crystal. The first wedge crystal separates the two beams, blocking the backward propagation of the signals into the input fiber.

3.4.1.1 Performance Capabilities and Limitations of Optical Isolators

A single-stage isolator [4] provides a minimum isolation of 32 dB, maximum insertion loss of 0.5 dB, return loss better than 60 dB, PDL significantly less than 0.05 dB, chromatic dispersion of less than 0.2 ps/unit length, instantaneous bandwidth of +/-20 nm and, typical power-handling capability of 300 mW. Single-stage isolators are widely used in optical transmission systems using distributed-feedback (DFB) lasers. Note the DFB laser frequency is most sensitive to back reflections that have adverse effects on the laser-gain profile and power stability. EDFAs with gain ranging from 25 to 40 dB are extremely sensitive to back reflections and often use more than one isolator to eliminate reflections from various joints and surfaces in the circuit. FO isolators have potential applications in telecommunications, biotechnology, FO gyros, and other optical systems requiring crosstalk-free operations.

In severe-reflection environments, multistage isolators will be found most attractive, but at the expense of higher cost, insertion loss, and complexity. A two-stage isolator offers isolation better than 55 dB, insertion loss not exceeding 0.75 dB, and PDL of less than 0.05 dB over the standard telecommunications wavelengths of 1310 to 1550 nm. A two-stage isolator can cost between $700 and $1500 in small quantity. Note FO isolators operating in the UV or near-IR regions are bulky and expensive. Isolators must be operated at power levels and temperatures recommended by the manufacturers to ensure stated optical performance and reliability.

3.4.2 Critical Performance Parameters of Circulators

Circulators are three-port devices that allow the use of a single optical fiber to transmit and receive optical signals with out compromising signal quality. These devices are generally used where reverse reflection signals are to be detected, compared, and measured with high accuracy. These circulators are available to operate at 1310 or 1550 nm wavelengths with one-meter-long pigtails. FO circulators are also available for four-port or eight-port operation. The PDL and polarization-mode dispersion (PMD) for circulators are higher than those for isolators due to device geometry and complexity. Commercially available circulators exhibit room-temperature insertion loss close to 1 dB, minimum isolation of 50 dB, typical bandwidth of 50 nm, maximum PDL of 0.2 dB, maximum PMD of 0.2 nm, and return loss better than 55 dB. High performance circulators are more expensive and bulky than isolators operating at the same wavelength. Typical circulator unit cost varies between $2800 and $4500, depending on the isolation, bandwidth, and other performance requirements. Circulators are preferred over isolators where high isolation and minimum insertion loss over wide spectral region are the requirements.

3.4.3 Critical Performance Parameters for FO Circulators

Insertion loss, PDL, return loss, and extinction ratio are the most critical performance requirements for high-performance circulators. These parameters must be given serious consideration prior to selection of FO circulators for specific system if measurement accuracy, optical system stability, and reliability are the principal requirements. The latest market survey [5] indicates that polarization-maintaining FO (PMFO) circulators are available for 1310 nm, 1550 nm, and other wavelengths. These devices exhibit a typical insertion loss of 0.5 dB, extinction ratio of better than 30 dB, and return loss close to 60 dB. These devices are available with various optical fibers. Special connectors are available for high-extinction-ratio, polarization-maintaining fibers. Fiber connectorization must be given top priority if stability and reliability in a compact in-line package are the critical requirements.

3.5 Optical Switches

Optical switches can be classified in various categories such as MEMS-based switches, AO switches, Rotman switches, and FO-based cross-matrix switches. Recent introduction of the IEEE 802.11 standard has played a key role in identifying suitable optical switches for wireless local-area networks (LANs), wireless telecommunications systems, and optical networking between various users. The 802.11 standard defines the physical (PHY) and medium-access-control (MAC) protocols for wireless LANs. The PHY protocol specifies three transmission options: one IR option and two radio frequency (RF) options (namely, direct-sequence spread spectrum [DSSS] and frequency-hopping spread spectrum [FHSS]). The PHY protocol uses two- or four-level Gaussian frequency-shift-keying (GFSK) modulation techniques capable of satisfying price and performance requirements in a broad range of network environments, plus additional features such as roaming and power management. All these options require the integration of cost-effective optical switches.

Optical switches are designed to connect desktop computers, notebook personal computers, and handheld electronic devices operating in a multifunctional network involving both the intra- and inter-building networks. Over the past decade, LANs have increased in speed from 1 Mbps to 1 Gbps as application and usage of these networks and electronic devices have gained tremendous popularity. New optical and electronic devices are evolving to bring network speeds close to 10 Gbps. Software need higher speeds and greater amounts of bandwidth because of more complex graphics used in applications such as MRIs, sophisticated tracking and detection algorithms required in military systems, and rapid transmission of digital/analog data. Effective handling of large amounts of data requires network infrastructures capable of supporting the bandwidth and switching requirements. New devices have been introduced that provide multiple functions. New products involving switching functions permit users to switch between two fiber networks or devices, providing operational flexibility with no compromise in efficiency or reliability. Fiber switching devices can also provide redundancy in FO-based installations, thereby significantly increasing the operational reliability of the system. In summary, FO-based switches offer several advantages, including higher speeds, long-distance operations, improved reliability under severe radio frequency interference (RFI) and nuclear radiation environments, and greater resistance to obsolescence and risk-free data transmission in critical applications.

3.5.1 FO-Based Crossover or Cross-Matrix Optical Switches

Crossover switches are also known as cross-matrix switches [3] and are widely used for optical switching applications because of their low cost and easy integration on a single substrate using SM optical fibers. A crossover optical switch architecture showing the integration of 2×2 switches on a single substrate is illustrated in Figure 3–5, while the deployment of such switches in a variable programmable delay line is shown in Figure 3–5.

Single-pole-single-throw (SPST), or 1×2, and double-pole-double-throw (DPDT), or 2×2, optical switches are widely employed by optical systems because of utmost simplicity and lowest cost. A DPDT switch has two input and two output ports and could operate either in a straight-through state (=) or in a crossover state ($\times$). Structural details and "OFF" and "ON" states of the switch are illustrated in Figure 3–6.

Two major categories of optical switches are available. One is based on a coherent coupling concept, while the other is based on a deflection phenomenon using either acousto-optic or electro-optic BRAGG interaction. Nevertheless, one must select an optical switch (OS) that is least sensitive to manufacturing tolerances and environmental factors such as pressure and temperature. An OS using a coherent coupling concept is based on modulating the mutual coupling between the two parallel channels. The architecture of a thin-film BRAGG switch is shown in Figure 3–6. Note the core diameter must be specified for minimum insertion loss and broadening effects over a wide spectral bandwidth.

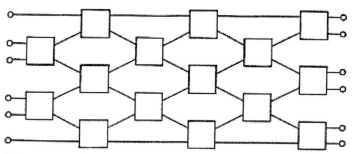

(a) Integration of 2x2 Single-Mode Optical Switches on a Single Substrate

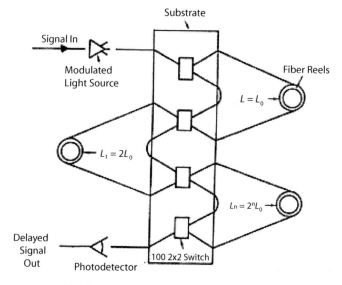

(b) Schematic Diagram of an Optical Variable Delay Line

Figure 3–5 *(a) Integration of optical switches and (b) schematic diagram of an optical variable delay line.*

3.5.1.1 Various Optical Switches for Telecommunications System Applications

A cross-connect optical switch (CCOS) is designed to switch optical signals between a pair of alternate output fibers of optimum numerical apertures. Fast optical switches promise to cure telecommunication's logjam problems. An 8×8 CCOS requires 64 switching nodes, a 16×16 CCOS requires 256 switching nodes, and a 32×32 CCOS requires 1024 switching nodes. Advanced optical switches are currently in development involving cross-connect configurations with

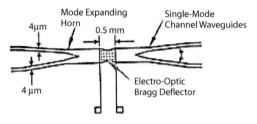

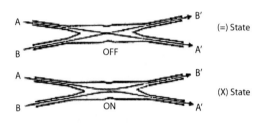

Figure 3–6 *(a) Architecture of a thin-film BRAGG optical switch and (b) "ON" and "OFF" switching modes.*

multiple input and output ports and add/drop switches, capable of switching individual wavelengths into and out of WDM signals. Various switching schemes are under aggressive research and development activities to meet diversified and complex switching requirements for future applications. They include optomechanical switches that move fibers or mirrors to redirect optical signals and solid-state switching devices capable of shifting light signals between optical fibers or planar waveguides. Active research efforts are focused on nonlinear switching devices for rapid wavelength conversion needed to switch optical signals from one wavelength to another, as in the case of a WDM system. High-speed optical logic gate (OLG) circuits are being developed for implementation in fast time-division-multiplexing (TDM) applications. OLGs use the optical signal contents to control the switching function. Designs for optican AND and INVERT logic gates operating at 100 Gbit/sec data rate have been optimized. The telecommunications industry is the major user of such optical switches. The rapid spread of WDM systems is creating an urgent need for an add/drop capability that can switch wavelengths into and out of a WDM signal. Dense-WDM systems are capable of carrying simultaneously dozens of wavelengths [6]. According to a published article, a dense-WDM system has demonstrated a transmitting capability of more than 80 wavelengths with high transmission efficiency. Telecommunications operators can add or drop certain wavelengths at system nodes without completely separating all wavelengths. Insertion loss, switching speed, and isolation between fibers or channels and cross talk are the most important performance specifications of an optical switch.

3.5.2 State-of-the-art (SOTA) FO-Based Switches

Recent advancements in FO and photonic technologies have played a key role in the development of SOTA FO switches. This switch combines three techniques: optical, photonic, and electronic. Because of complex design issues involved, SOTA switches are designed specifically for 1310 and 1550 nm operating wavelengths. The FO switch with A, B, and C interface uses a switch/converter (S/C) design configuration, as shown in Figure 3–7.

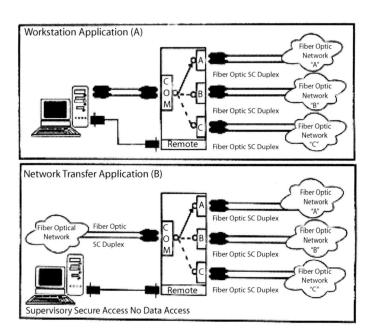

Figure 3–7 *Architecture of a SOTA switch for two distinct applications: (a) workstation application and (b) Network Transfer application (courtesy of Electro Standards Laboratory, Cranston, Rhode Island)*

This switch has a multiple duplex fiber connector interface on the A, B, C and COMMON ports. This SOTA switch offers two potential applications, namely, workstation and network transfer between three other optical devices or networks. The built-in circuits within the switch assembly offer isolation better than 50 dB between the A, B, and C devices. This switch is best suited for fallback network or FO link backup and sharing applications. A push-button switch on the front panel permits selection of the devices or networks. Red LEDs display switch position and indicate the electrical power presence. The FO provides an interlace for a specific wavelength and data speed from 100 Mbps to 1 Gbps. The switch architecture includes an RS-232 interface capability that provides enhanced a supervisor remote port (SRP). Upon proper authentication or identification, a computer in terminal mode connected to this port can communicate with the unit to determine the operating status, change the switch position as desired, or lock out front-panel switch function. A modem can be connected to the SRP port for remote access to the SOTA switch. Note access to the SRP is only available upon the verification of a password. Optical graphical user-interface software is available for use with a computer running Windows 95/98, NT, Me, and 2000-XP. This switch can be designed for multiple configurations such as electronic relay, routing, power reboot, fast Ethernet, rack mount, and desktop. The SOTA switch can interface both the local and remotely controlled networks and offers adequate isolation and crosstalk between devices or networks.

3.5.3 High-Performance Ethernet Switch

A memory-based Ethernet switch fabric on a chip is capable of incorporating a full dozen 10 Gbit/sec Ethernet ports. The solid-state chip reduces the port cost by two orders of magnitude. The switch architecture includes a high-speed buffer memory, high-speed I/O macrocells, and a high-performance multiported memory capable of handling simultaneously read and write operations to and from all one dozen 10 Gbit/sec ports. With all ports active, the chip can offer an aggregated bandwidth of 240 Gbit/sec. The multipart memory meets the throughput requirements, even with minimum-size Ethernet packets. This switch can satisfy the short-latency and high-throughput packet-switching requirements [7] of high-end servers and advanced storage systems.

3.5.4 MEMS-Based Optical Switch

All-optical switching offers several benefits such as cost, compact size, and minimum power consumption. An MEMS-based optical switch provides large-scale photon cross-connects leading to the architecture of a 288×288 channel optical switch. The three-dimensional (3-D) MEMS switch offers design flexibility with minimum cost and complexity. The optical signal enters via an input low-loss fiber and is directed to one of the MEMS mirrors through a lens array. The mirror reflects the optical signal to another MEMS mirror that reflects it again to the output lens array and to a proper output optical fiber. Since the MEMS mirrors rotate on two axes, they can move the signal in three dimensions.

Maintaining a low, stable, and uniform insertion loss is critical, especially for optical networks in which the signal encounters multiple switching nodes and fiber spans [8]. This type of switch has demonstrated an insertion loss of 1.4 dB over the spectral region from 1260 to 1625 nm. Precision positioning of optical fibers is necessary to avoid bending loss at longer wavelengths. In addition to smaller footprints and lower power consumption, the MEMS-based switches enable network providers to eliminate lasers, receivers, transponders, multiplexes, and redundant electronic switch ports at transit nodes, resulting in significant reduction in power consumption and operating costs. Because MEMS-based switches are bit-rate, protocol- and wavelength-independent, they can support any combination of transport formats and 1, 10, and 40 Gbit/sec signals with minimum cost and complexity. Matrix optical switches with 128×128 and 256×256 element configuration are readily available.

3.5.5 Fiber-Laser-Based Q-switch

Fiber lasers are becoming most attractive for commercial and industrial applications because of their compact size, rugged packing, higher efficiency, and improved reliability with passive cooling and high beam quality [9]. This switch uses a stimulated Brillion scattering technique, illustrated in Figure 3–8, that generates a narrow seed pulse capable of traveling through the population inversion and quickly gaining amplitude. The width of the seed pulse is dependent on the interaction between the acoustic wave and the optical fiber core. This pulse width typically is close to a nanosecond for a core diameter of 6 microns.

Brillion scattering is considered a feedback mechanism that dramatically increases the Q of the resonator during the nanosecond lifetime of the acoustic wave. The stimulated Brillion scattering thus serves as a passive Q-switch within the fiber laser resonator. An ytterbium-doped, double-clad fiber with a length ranging from 15 to 20 meters is used, featuring an inner cladding with a D-shaped cross-section to suppress doughnut modes that do not couple optical energy to the core. A diode laser is used as a pumping source capable of providing both the continuous wave (CW) and pulse-operating modes, as illustrated in Figure 3–8. Under continuous pumping, the fiber laser reaches a threshold around 560 mw and begins random pulsing at about 800 mw. The laser pulses generated by Brillouin backscattering are very short (less than 10 ns) and are very unstable. Stabilization of the laser pulses is accomplished by switching the pump diode to pulsed-mode operation. Under pulsed-mode operation, a peak power exceeding 100 kW can be obtained. This is an order of magnitude greater than power available from a Q-switched conventional fiber laser. The bandwidth of the laser can be reduced to 0.04 nm by a slight change in resonator configuration. This fiber laser can be tuned over a 5.5% bandwidth in the spectral region from 1080 to 1140 nm.

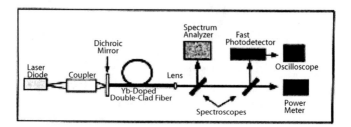

Figure 3–8 *Stimulated Brillouin scattering concept to design a Q-switch with a dode-pumped Yb-doped double-clad fiber label capable of producing a peak power an order of magnitude greater than a conventional Q-switch fiber laser.*

3.6 High-Speed Optical Interconnect (HSOI) Device

Integration of very large signal integration (VLSI) and photon technologies offers a high-speed optical interconnect device capable of achieving a thousand times the speed per unit power. Based on Moore's law, a single-chip processor can achieve a speed better than tens of teraflop, but the throughput of the electronic interconnect will be in the 1-to-10 Gbit/sec range. In brief, the electronic interconnect technology currently being used will not meet processing requirements exceeding 10 teraflops.

The HSOI device architecture requires the use of a vertical-cavity surface-emitter laser (VCSEL) diode array, photo detector for I/O circuits, free-space optics for interconnects paths, and wide-bandwidth interface optical fibers coupled with CMOS technology. Optics have a distinct advantage when it comes to large numbers of direct interconnects between the processors because light beams can intersect without crosswalk. In addition, optical interconnect can be scaled up with very little increase in power dissipation. These HSOI devices provide architectures [10] leading to all-optical crossbar switches and multistage switching networks, which have potential applications in complex military radar systems.

3.7 FO Bundle (FOB)

An FO bundle (FOB) is nothing but a low-loss FO cable consisting of several optical fibers, which may include silica/silica, plastic-clad silica, and hard polymer-clad silicon fibers. Several bundle configurations are possible including single, bifurcated, trifurcated, and multilegged with wide values of NAs. The majority of FOBs are made from all silica fibers of various NA values operating from 160 to 2500 nm, from plastic-clad silica fibers with NA ranging from 0.3 to 0.4 operating from 200 to 2400 nm, or from hard-polymer-clad silica fibers with NA ranging from 0.37 to 0.48 operating from 200 to 2200 nm. Various jacket options are available for FOBs depending on the operating environments. Note these FOBs are resistant to radiation and can operate over wide temperature ranges without compromising performance or reliability. Their potential applications include industrial sensors, laser delivery, smoke detection, nuclear plasma diagnostics, UV photolithography, photodynamic therapy (PDT), chemical analysis, remote sensors operating under harsh thermal and toxic environments, and complex military systems. In summary, FOBs are best suited for commercial, industrial, and space and military applications where large optical energy is to be delivered with high reliability in harsh operating environments.

3.7.1 Polarization-Maintaining Capability

The polarization-maintaining (PM) capability of these bundles is critical. The FOB must deploy SM optical fibers that preserve the polarization of the optical signal as it travels through the fiber length. Sophisticated assembling techniques and customized testing procedures provide precise alignment in both the mechanical and optical planes. FOBs with PM capability have demonstrated low insertion losses and stability can be terminated with industry-standard optical connectors for use at 1330 or 1550 nm operating wavelengths. FO connectors such as 8308-SC and 8208-FC exhibit insertion loss of less than 0.15 dB over the temperature range of $-40°$ C to $+100°$ C. These connectors use zirconium-ceramic ferrules, which when polished can provide reflectivity better than -55 dB.

3.8 FO Probes

FO probes are widely used in pharmaceuticals, medical diagnostics, biomedical research, optical signal analyzers, and auto-engine-performance analysis. The FO probe can be connected to an analyzer for simultaneous data collection in situ from a parallel-synthesis automated label reactor used in drug-recovery applications. This probe can be used with noncontact and immersion optics for analyzing powers, slurries, and liquids in a laboratory or under process environments. Flanged process optics for coupling to extruders at a pilot plant or manufacturing location is preferred to ensure high process efficiency and improved quality control.

3.8.1 FO Raman Probes for Spectrographic Applications

For spectrographic applications, FO Raman probes are best suited and are readily available in 785, 633 and 532 nm wavelengths. The FO Raman probe incorporates hard oxide-coated filters and antireflection-coated (AR-coated) lenses. The filters and AR-coated lenses provide protection from laser light. This probe can be used with a compact sample holder for accurate measurements of liquids and solids. The probe can be designed for use in harsh thermal and mechanical environments.

3.8.2 FO Probes for Engine Diagnosis

High-temperature FO probes are widely used by automobile manufacturers to monitor engine performance under severe thermal and mechanical environments. The auto industry uses FO probes to collect important performance data from various sensors that are used to reduce the harmful emissions from gasoline-powered engines. An FO sensor is integrated with a spark plug to provide real-time cylinder pressure information. FO probe detection sensitivity as high as 0.0034 red/sec can be achieved using an intrinsic optical fiber Fairy-Perot interferometer. Monitoring the engine's internal pressure and using a semiconductor laser emitting near the 1310 nm wavelength, one can measure the variations in the reflectance of the metal-coated fiber. Optical fibers capable of withstanding harsh thermal and mechanical environments are required for the above applications.

3.8.3 FO Probes for Oscilloscopes

Probes using high-performance optical fibers are required to display electronic data on oscilloscopes. Such probes are also known as optical-to-electrical converters because the input to the probe is optical, while the output of the probe is electrical and directly connected to the oscilloscope input terminal. Optical fibers with ultralow insertion loss, minimum bending loss, and lowest dispersion are recommended to provide accurate measurements. The optical-to-electrical converter consists of an FO cable with minimum insertion loss and dispersion, FO connectors with low loss, and a sensitive optical detector optimized for a specific spectral region and bandwidth. An FO connector-detector designated as ST-silicon is used for a spectral region of 400 to 1000 nm, as ST-InGaAs for a spectral region of 900 to 1700 nm, and as FC-InGaAs for a spectral region of 900 to 1700 nm and 750 MHz radio frequency (RF) bandwidth. Note that the converter cost increases with the increase in detector sensitivity and RF bandwidth.

3.9 FO-Based Optical-Power Combiners

FO-based optical-power combiners (OPCs) are also known as mode multiplexing optical couplers. These optical combiners combine emissions from several solid-state diode lasers with combining efficiencies as high as 98%. Typical yield rates are in excess of 80%, and the average insertion loss is less than 0.4 dB [11]. Essentially, an OPC combines the power outputs from several diode laser sources into a single output fiber. Today's commercially available fiber-coupled multimode pump lasers are capable of delivering an optical CW power level around 2 W, while the output of the fiber-coupled bar diodes exceeds 120 W, with a potential power output exceeding 200 W in the near future.

3.9.1 Fabrication Techniques for OPCs

Well-proven fabrication techniques must be used to avoid mode suppression and excessive insertion loss problems. Fabrication of an OPC for diode-pumped laser systems requires careful bundling of optical fibers, heating and pulling them lengthwise to avoid any abrupt fiber break. Such a combiner merges emissions into a multimode output fiber element known as a double-clad fiber (DCF). This DCF is widely used for high-power optical amplifiers and fiber lasers. The combiner configuration for an amplifier involves an SM fiber at the center of the combiner and a tapered section. The tapered section of the combiner is cut after being formed by a fusion process and then is sliced to a DCF that has a rare-earth-doped core. The SM fiber must be well coupled to the core of the DCF to ensure efficient coupling of the light from the diode emitters to the MM fibers and to maximize optical power throughput between the fibers and the fused region of the combiner. In addition, the output fibers must have sufficient capacity to accommodate the intensity of the combined beams with minimum distortion.

3.9.2 Brightness of the Combiner

Conservation of the brightness is based on the Lagrange invariant of an optical system and is generally characterized by a quantity called etendue (E), which is the product of the area of a light beam and the solid angle subtended by the optical beam. An increase in brightness implies reduction in etendue. Etendue for a step-index MM optical fiber can be written as

$$E = [(\pi)^2 / 4(NA)^2 (D)^2] \qquad 3.1$$

where NA is the effective numerical aperture and D is the diameter.

Readers are advised not to confuse brightness units for other sources. Airglow or aurora brightness is expressed in "Rayleighs," celestial brightness is in "Steller," and display brightness is in "Foot-Lamberts."

If the step-index fiber is tapered, its E remains constant and the effective NA increases as the D decreases. The NA parameter refers to the maximum angle of light entering or exiting the optical fiber. Two fundamental requirements need to be satisfied to transfer optical power efficiently between two optical segments or elements. First, the E on the input side must be equal to or less than that on the output side. Second, the cross-section areas of the input and output fibers must match at the junction. In summary, actual design requirements for a high-power optical combiner are application specific.

3.9.3 Combiner Requirements for Various Power Levels

Design requirements for a combiner are dependent on the output power level and the number of optical sources involved. Low-power applications, such as combining fiber lasers for Raman pumping, requires combiners to carry power levels that are typically less than 10 W. The same combiner can be updated for higher-power industrial applications where the optical power may exceed 40 W. Combiners with 40 W power ratings do not require sophisticated cooling.

Twelve combiners consisting of twenty identical step-index input fibers can meet the following performance requirement [11]:

- Core diameter: 105 microns
- Cladding diameter: 125 microns
- Protecting coating diameter: 250 microns
- NA: 0.22

Each step-index output fiber will have the following dimensions:

- Core diameter: 200 micron
- Cladding diameter: 220 micron
- Protecting coating diameter: 500 micron
- NA: 0.48.

3.9.4 Deployment of OPCs in Military Applications

A high-power OPC has potential applications in military systems. These applications may include airborne target illumination by attack aircraft, airborne laser tracking sensors, and laser range finders. OPCs for military applications must meet stringent shock, vibration, and thermal requirements. Preliminary studies performed by the author indicate that coherent combination of hundreds of fiber lasers can generate peak optical power levels exceeding 100 kW. The studies further indicate that a moderate power level greater than 1 kW is possible through coherent combination of ten fiber lasers, which is adequate for short-range laser range finders.

3.10 FO Tunable Dispersion Compensators

Chromatic dispersion can degrade the digital signal traveling through an optical fiber due to a pulse widening effect and start overlapping. Various techniques can compensate for chromatic dispersion, but FO-based compensation techniques offer cost-effective solutions. Furthermore, tunable FO-based dispersion-compensation techniques are best suited for WDM signal applications regardless of channel count, channel spacing, or data rate.

In general, there are two possible approaches to solve the dispersion problem: canalized and continuous solutions. In the case of the canalized approach, the channels of the WDM signals are separated and each channel is compensated individually. In the case of the continuous solution, chirped Bragg gratings are used, where each channel has its own grating. The continuous solution is dependent on the length of the dispersion-compensation fiber whose dispersion has the opposite sign to that of the normal fiber. The Fourier components of the signal that are faster in the normal fiber are slower in the compensating fiber. In this way, all signal components or parameters take the same time to travel through both fibers. This approach is continuous because it works for all WDM channels carrying secured communication data and does not require channel separation. The tunable dispersion compensator shown in Figure 3–9 can operate over a broad band of optical frequencies. This technique relies heavily on adjustable propagation through higher-order mode fibers depicted in Figure 3–9.

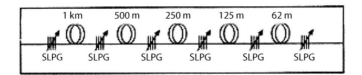

(a) Tunable Dispersion Compensator using Five Lengths of Fiber and 6-Switchable Long-Period Fiber Gratings (SLPGs)

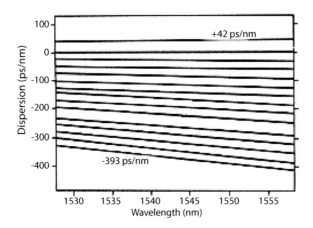

(b) Lines Showing the Dispersion as a Function of Wavelength

Figure 3–9 *(a) Architecture of a tunable dispersion compensator and (b) dispersion as a function of wavelength for two polarization modes propagating in fibers following grating.*

The optical signal can propagate through the fiber in either LP_{01} or LP_{02} modes using switchable, long-periodic fiber grating (SLPFG) in front of each channel. The length of each fiber determines which mode will propagate in that length of fiber. Furthermore, each grating can be switched to determine which of the two polarization modes propagates in the fiber following that particular grating. The total dispersion can be adjusted by switching the gratings because the two modes have different dispersion values [12]. Assuming a system consisting of five gratings, 32 (2^5) different values of total dispersion are possible; samples of these values are shown in Figure 3–9. The bandwidth of this approach is close to 32 nm because it is limited by the bandwidth of the long-periodic gratings (LPGs). These gratings with wide bandwidth can be used to improve the compensator bandwidth to cover the C-band spectral region. The insertion loss is strictly due to the fiber splices and can be significantly reduced by improving the quality of the splices.

3.11 FO-Based Light Scopes

FO-based light scopes, also known as fiberscopes, consist of a water-resistant optical, built-in light source and a focusing mechanism to adjust focus distance from infinity to a lowest distance of 1 cm. The insertion end of the scope has a typical diameter of 8 mm at the tip and 10 mm with side-view adaptor attached that allows the light into crooked locations or inspection sites difficult to reach. The effective length varies from 100 to 300 cm, which allows visual inspection of the far-back locations. The effective length provides optimum flexibility for inspection in crooked pipes and other difficult-to-inspect locations. The built-in light source provides high portability and brightness levels as high as 19,000 lux. This is sufficient to examine hidden objects at viewing angles up to 27 degrees. The fiberscope is considered an excellent industrial or military endoscope, capable of providing testing and visual inspections for automobiles, complex industrial machinery, jet engines, turbines, compressors, petrochemical plants, and vessels.

The light scopes can be integrated with digital or CCD cameras, PC cards, laptops, and desktops for comprehensive inspection of the critical elements or parts of a complex system and simultaneous recording and display of test data in real time. The FO scope plays a key role in internal inspection of automobile engines, transmission gears, differentials, and air conditioners. In summary, the FO scope allows instant inspection of the critical areas or surfaces of turbine or compressor blades, combustion chamber interiors, rocket engines, clogged-toilet or plumbing problems, boiler interiors, and a host of other industrial, military, and space systems.

3.12 FO Strain Sensors

FO strain sensors represent a product integration of photonic and FO technologies. This photonic solution enables monitoring of strain fatigue in aircraft, missiles, or spacecraft under simulated operating environments. Replacement of traditional electronic sensors with optical fibers has not only eliminated weight and space constraints, but also EMI problems during the strain tests. As an aircraft or spacecraft ages, decompression of the cabin can weaken the structure, deicing agents can corrode the metal joints and vibration can seriously compromise the mechanical integrity of the joints, and control surfaces. Testing of critical aircraft-control surfaces can provide evidence of fatigue or cracks using FO strain sensors with minimum cost and complexity.

The FO strain sensor deploys the optical frequency domain-in reflectometry techniques, which permit spatially distributed measurements from hundreds of fiber Bragg grating sensors along a single optical fiber. The fiber is glued to a designated aircraft surface and is capable of monitoring the refractive-index change as the surface experiences tension or compression. Each sensor acts as an independent interferometer with a beat frequency proportional to its length along the fiber axis. A tunable laser is used to interrogate the sensors over a 7-nm spectral range at 60 nm/sec tuning rate, which corresponds to a strain range of about +/-3500 microns over a scan period of 120 milliseconds. The reflected light is then detected, demodulated, and analyzed. Each sensor has a unique spatial position and its individual spectra. The data collected are displayed as strain field maps that indicate or identify the location of faults and fatigue in a spectral region. These FO strain sensors can be installed on critical surfaces of a military supersonic jet aircraft or spacecraft for continuous in-flight monitoring of the strain data.

3.13 FO-Based Security Uniforms

Deployment of FO-based security uniforms will provide battlefield solders effective protection against ballistic, chemical, and biological threats [12]. The FO technology, when inserted in solders' uniforms, can help in identifying friend from foe and can administer medical treatment in the case of an emergency on the battlefield. The futuristic battlefield suit will be a lightweight system that is capable of integrating several functions into a material called photonic bandgap (PBG) fiber. When PBG fibers are integrated into the fabric used in the battlefield suit, the uniform reflects a specific wavelength in the IR region that would be recognized by fellow solders in the brigade or company. The mirrored fabric will protect soldiers from radiation and can change to camouflage, which in turn will provide effective protection. The research scientists at MIT, working in the nanotechnology area, have developed this particular fiber with hundreds of alternating layers that are capable of tuning over various IR wavelengths.

The material used in the security uniform combines the characteristics of metallic and dielectric mirrors that reflect light from every angle. The photonic material must be ductile, similar to a polymeric fiber. In addition, durability and stability are the principal requirements because uniforms must tolerate the stresses of being worn and laundered in battlefield environments. Furthermore, uniform fabric must be mass produced at a reasonable and affordable cost. The FO-based uniforms have potential applications for emergency workers, police, firefighters and others requiring similar protection under harsh field environments.

3.14 FO-Based Interrogation Sensors

FO-based interrogation sensors are best suited for military applications requiring important information from enemy soldiers or army deserters. The FO-based interrogation sensor employs the latest photonic fiber technology. This sensor measures the strain, stress, and temperature of the fiber Bragg gratings (FBGs) in an optical fiber. The sensor transmits and receives the light from a compact interrogation unit via an optical fiber with sensor gratings. The fiber can be embedded or attached to any structure. The compact optical fiber sensors do not require electrical power, even for extended operation. The system is also suitable for monitoring and recording the structural health of tunnels, building structures, and bridges, and for medical diagnostic applications. The fiber gratings and software interface can be tailored for specific applications.

An FBG provides the most efficient form of filter when it reflects a single wavelength and transmits the other wavelength signals. In an FBG, the variation in refractive index is continuous and much smaller. The smaller the refractive-index variation, the higher the interrogation reliability and accuracy of the sensor will be. The sensor reliability and accuracy are the most vital parameters of this particular sensor.

3.15 FO Lighting System

The latest research and development activities have developed a unique lighting system based on FO technology. This system uses a woven FO technology that enhances back lighting. Standard FO backlighting panels can be illuminated by five or six different color lead emitting diodes (LEDs). These panels can also be lit with incandescent and halogen light sources, if higher brightness is the principal requirement. Sources used by this system require a current as low as 10 to 20 mA at 2 VDC. FO panels do not add heat or EMI to interfere with switching operation. Optical fibers used in the system are not affected by extreme humidity and temperature. This system can enhance the back lighting of liquid crystal displays (LCDs), keypads, desktops, laptops, and control panels used for machine vision applications. Major advantages of the lighting system include higher brightness, durability, compactness, longer life, minimum power consumption, and availability of a wide range of colors. This system offers a low profile, maximum design flexibility, very low startup costs, compact packaging, and no maintenance.

3.16 FO Collimators

Miniaturized dual-fiber collimators can fall into three distinct categories: transmission, reflection, and crossing types. The collimators have an epoxy-free optical path with minimum loss and a collimator diameter not exceeding 1.3 mm or 1300 microns. The fiber collimators provide low insertion loss, high return loss, and reliable performance over extended operating periods. These collimators are best suited for optical-component integration. The collimators have potential applications in miniature optical switches, variable optical attenuators, and coarse WDM communication devices.

3.17 FO Polarizers and Depolarizers

FO depolarizers are passive devices that typically provide a degree of polarization of less than 5%, insertion loss not exceeding 1 dB, and zero-back reflection over a wide spectral range. The depolarizers, which are designed to match the source spectrum, use the coherent properties of the optical source to randomize the state of polarization. High-performance, all-fiber depolarizers have been developed by several photonic companies.

Evanescent field FO polarizers are produced by replacing the cladding in the locally processed region of the optical fiber with an embedded polarization selective material. Within the polarization region, one polarization mode of the SM fiber is highly attenuated, whereas the other mode propagates virtually with no insertion loss. Extinction ratios better than 55 dB are readily achievable while maintaining extremely low transmission loss of the required polarization mode.

Reduced-cladding polarizing devices are broadband but can be optimized for 1300 nm and 1550 nm operating wavelengths. Unique performance parameters such as low transmission loss that approaches zero reflection and high extinction ratios make these devices most attractive for polarization-sensitive sensor applications, such as the optical fiber gyroscopes and polarimetric fiber systems. These devices offer wide spectral bandwidth, high reliability, rugged packaging, and minimum cost. Optimum performance is available over the 1530-to-1570 nm and 1280-to-1320 nm spectral regions.

Polarizers can be tested and graded into two performance groups: SM standard fiber and polarization-maintaining (PM) fiber. Typical insertion loss is about 0.2 dB for SM/SM series and about 1 dB for PM/PM devices, excluding the optical connector losses. All-fiber device technology offers return loss better than 70 dB.

3.18 Erbium-Doped Microfiber (EDMF) Laser

An EDMF laser is a low-cost, high-power, compact laser module and is best suited for sensing, testing, and measurement applications. The telecommunication-grade thermo-electrically (TE) cooled, SM diode laser pumps the EDMF laser, which is capable of providing a peak output power in excess of 100 mw over the telecommunication C-band (1530–1565 nm). This unit offers high reliability while operating under harsh operating environments. The outstanding features of an EDMF laser include very narrow line width, a single longitudinal mode, single polarization, high wavelength stability, an integrated pump laser, microprocessor control, reduced noise level, an improved polarization extinction ratio, and low power consumption. Its potential applications are sensor testing, LIDAR, and telecommunications equipment.

3.19 FO Polarization Scramblers

As the bit rate increases, FO communications systems become increasingly sensitive to polarization-related impairments or issues. Significant problems can arise related to polarization mode dispersion in optical fibers, PDL in passive optical components in the system, polarization-dependent modulation in electro-optical modulators, polarization-dependent gain in optical amplifiers, polarization-dependent center wavelength shift in WDM filters, polarization-dependent receiver response, and polarization-dependent sensitivity in sensors and coherent communications systems. In brief, higher bit rates are making the polarization property an important issue in the design of high-speed optical communications networks or systems.

3.19.1 Basic Scrambling Principles

A polarization-scrambling technique can play a key role in mitigating most of the polarization-dependent related problems or issues stated above. Scrambling occurs when the state of polarization of fully polarized light varies randomly at a relatively low rate. Although the state is well defined and the degree of polarization is close to 100% at any instant, the degree of polarization will approach zero. Therefore, the degree of polarization for scrambled light depends on the average time or the detection bandwidth of the observer [13]. Polarization-scrambler activity changes the state of polarization using the polarization-modulation technique.

3.19.2 Scrambler Types

Different scrambling technologies are available, such as lithium $LiNbO_3$, resonant fiber-coil, and fiber-squeezer systems. $LiNbO_3$ scramblers (see Figure 3–10) deploy the electro-optic effect to modulate the state of polarization. The resultant modulator speed is affected by high insertion loss, PDL, residual amplitude modulation loss, sensitivity to input polarization state, and cost of the device.

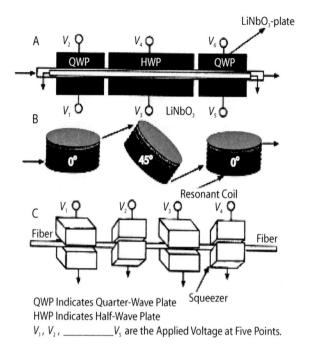

Figure 3–10 *Polarization scrambler devices based on (a) LiNbO₃, (b) on-fiber resonant coils, and (c) and on-fiber squeezers.*

Multiple modulation sections with different electric field directions could make the device less polarization sensitive, but at the expense of additional cost and complexity. Trade-off studies must be performed to determine the various critical parameters that contribute higher costs.

A scrambler based on resonant fiber coil is designed by winding optical fiber around an expandable piezoelectric cylinder of appropriate dimensions. When an electric field is applied, it causes the cylinder to expand. That in turn induces birefringence in the optical fiber via the photoelastic effect, which is dependent on the photoelastic coefficients of the piezoelectric material used. If the frequency of the electric field is in resonance with the piezoelectric cylinder, the induced birefringence will be large enough to cause sufficient polarization modulation, even with a low applied electric field.

Multiple fiber cylinders can be cascaded with different orientations to minimize polarization sensitivity. This alternative design concept has the advantages of low insertion loss, PDL, and system cost. However, this design suffers from large size, low scramble speed, and high residual phase modulation due to significant fiber stretching when the fiber coil expands.

A fiber-squeezing technique is capable of inducing enough birefringence to cause large polarization modulation if the input polarization is 45 degrees from the squeezing axis. Cascading several fiber squeezers oriented 45 degrees from each other can produce a polarization-insensitive scrambler. The cascaded device can operate resonantly at higher scrambling frequencies or nonresonantly at lower frequencies.

3.19.3 Performance Comparison of Various Scrambling Devices

A $LiNbO_3$ scrambler offers several benefits, such as low insertion loss, PDL, and development cost. When compared with a fiber-coil scrambler, this device offers maximum scrambling flexibility and small size. In addition, it has an edge over other scramblers both in terms of low residual phase modulation and residual amplitude modulation. Low residual phase modulation is important for avoiding interference-related noise, whereas low residual amplitude modulation is critical when the scramblers are used for PDL and degree-of-polarization measurements of optical devices or systems.

3.19.4 Elimination of Polarization Fading

Scramblers are best suited for eliminating the polarization fading of a fiber sensor. Placing a scrambler in front of a polarization-sensitive instrument, such as a diffraction-grating-based optical spectrum analyzer, can also effectively eliminate its polarization dependence if the scrambling rate is sufficiently faster than the detector response used by the instrument. A scrambler can help in the measurement of the PDL using a digital scope.

The PDL can be calculated using the following expression:

$$\text{PDL} = 10 log(V_{max}/V_{min}) \qquad 3.2$$

where V_{max} and V_{min} are the *max*imum and *min*imum signals.

Raman amplifiers generally exhibit strong polarization-dependent gain variation if the pump laser is highly polarized. Minimizing gain variations requires a depolarized pump source. The degree of polarization of the pump source determines the polarization-dependent gain of the amplifier. Scramblers are also used for accurate measurement of the degree of polarization on a digital scope.

3.19.5 Potential Applications of Scramblers

As stated earlier, polarization scramblers have numerous applications in optical communications networks, FO sensor systems, and test and measurement systems. Deployment of such a device at the transmitter side in an ultralong-haul communications system minimizes polarization-dependent gain variation or polarization hole burning of an EDFA. For this particular application, the scrambling rate must be significantly faster than the inverse of the gain recovery time constant of the EDFA. Scramblers will be found most attractive in the monitoring of polarization mode dispersion in a WDM telecommunication system.

3.20 Summary

This chapter summarizes the performance capabilities and limitations of FO-based passive components and devices widely used in ultralong-haul communications systems, commercial equipment, industrial sensors, and scientific research projects. Performance parameters are identified for FO devices such as directional couplers, attenuators, fiber lasers, variable-ratio PM couplers, isolators, circulators, switches, polarization scramblers, frequency select filters, and PDL compensators. Critical performance requirements are discussed for the FO components widely used in WDM and dense-WDM telecommunications systems (see Figure 3–1). Some of these FO devices, which play a critical role in electro-optic, optoelectronic, photonic, and IR sensors, are discussed in greater detail. Several types of optical switches are described, with emphasis on switching speed and insertion loss. Tunable dispersion compensators are briefly discussed with emphasis on channel separation. Requirements for FO–based strain sensors to monitor the structure fatigue are identified. Benefits of FO-based security uniforms capable of providing effective protection against ballistic, chemical, and biological threats are summarized. Performance requirements for polarizers, depolarizers, and polarization scramblers used in various applications are also discussed.

3.21 References

1. Jha, A. R. (2000). *Infrared technology: Applications to electro-optics, photonic devices, and sensors* (p. 208). New York: John Wiley and Sons, Inc.

2. Jha, A. R. *Acousto-optic tunable filters* (technical proposal, pp.12–16). Technical Proposal, Cerritos, CA: Jha Technical Consulting Services.

3. Jha, A. R. (1994, July). *Programmable microwave fiber-optic delay line* (technical report, pp. 6–12). Cerritos, CA: Jha Technical Consulting Services.

4. Guest Editor, (1998, November). Isolators protect fiber optic lasers and optical amplifiers. *Laser Focus World*, 148.

5. Editor, (2002, October). Fiber optic circulators. *Photonic Spectra*, 149.

6. Contributing Editor, (1998, September). Optical switching promotes cure for telecommunications applications. *Laser Focus World*, 69–71.

7. Bursky, D. (2003, July). High performance, 10-Gbit/sec ethernet switch knocks price down. *Electronic Design,* 44.
8. Boch, G. (2002, October). Optical switch offers 1.46 dB loss. *Photonic Spectra,* 100.
9. Hitz, B. (2003, June). Fiber laser Q-switched by stimulated Brillouin scattering. *Photonic Spectra,* 100.
10. Brown, C. (2003, May). High-speed optical interconnect in the works at DARPA. *Electronic Engineering Times,* 57.
11. Barron, C., & Lindsay, K. (2002, January). Multimode power combiners pump up. *Photonic Spectra,* 153–158.
12. Johnson, B. D. (2003, March). Security uniform. *Photonic Spectra,* 37–38.
13. Yao, S., et al. (2003, April). Scramblers to polarization related impairments. *Photonic Spectra,* 90–95.

CHAPTER 4

Fiber Optic Components for Military and Space Applications

This chapter defines the performance requirements for FO components and devices for potential applications in military and space systems. Major emphasis will be placed on FO components best suited for military-systems applications, such as IR missiles, drones, phased array antennas, transmit/receive modules for deployment in missile defense radars, programmable optical delay lines for integration in electronic countermeasures (ECM) equipment, unmanned air vehicles, battlefield sensors, and laser tracking and illumination systems. In addition, state-of-the art requirements for the special FO devices which will provide significant performance improvement in some critical military sensors will be briefly discussed. Such components or devices include fiber Bragg gratings (FBGs), tunable dispersion compensators, fiber lasers, FO gyros and erbium doped fiber amplifiers (EDFAs).

4.1 FO Delay Lines for ECM Equipment

Delay lines play a key role in the design and development of range-gate-pull-off (RGPO) systems. Note RGPO is known as a deceptive technique that provides effective ECM against terminal guidance seekers. This technique plays a critical role in countering the threats posed by hostile missiles to defensive and offensive weapon systems. When coupled with various angle-tracking deception techniques, the RGPO technique can counter threats posed by both the CW and pulsed Doppler seekers. A variable or programmable delay line is the heart of a RGPO system as well as a critical element of the ECM equipment.

4.1.1 Potential Delay Line Types and Design Parameters

Potential delay line types and design parameters will be briefly discussed in terms of performance capabilities and limitations, cost, and complexity. Delay elements can be made from wave guide, copper microstrip, coaxial line, and acoustic and FO transmission lines. The trade-off studies performed by the author [1] reveal that a delay medium using copper microstrip or a surface acoustic wave (SAW) device or Niobium transmission line suffers from high propagation losses compared to that from a FO-based design, as shown in Figure 4–1. The trade-off studies further reveal that the operating bandwidth and the amount of delay are dependent on the optical-source wavelength, as illustrated in Figure 4–2. Curves shown in Figure 4–2 indicate that a source wavelength of 1.55 microns offers maximum bandwidth using an SM optical fiber (SMOF) with minimum length. Investigation of various transmission lines indicates that an FO delay line offers widest RF bandwidth, minimum weight, lowest insertion loss and compact packaging over a wide operating temperature range with no compromise in performance.

Important characteristics of various delay structures are summarized in Table 4–1.

Table 4–1 *Characteristics of various delay-line structures.*

Characteristic	Structure for 200 ns Delay Line			
	Wave Guide	Coaxial	Acoustic	Fiber Optic
Dispersion	Low	High	Low	Negligible
Bandwidth	Octave	Multioctave	Half-octave	Octave
Volume (cu. in)	1300	200	< 1.0	< 0.25
Insertion loss (dB)	19	45	28	0.25
Weight (lb)	15	8	0.5	0.33
Random Noise Capture	Serious	Serious	Medium	Minimum

Delay lines using other structures suffer from excessive weight, large size, high insertion loss, large dispersion, and large delay variations. They are also vulnerable to electronic jamming and nuclear radiation. An FO delay line using a low-loss SMOF will overcome most of the problems associated with other types of delay lines. It is evident from Table 4–1 that the insertion loss in an optical fiber structure is independent of RF and is less than 0.25 dB for one microsecond delay in a SMOF. An SMOF is the only transmission line that can offer very large delay with minimum loss, weight, and size.

4.1.1.1 Digitally Programmable Optical Delay Line

A digitally programmable optical delay line (DPODL) is best suited for ECM systems requiring maximum design flexibility to meet changing environmental

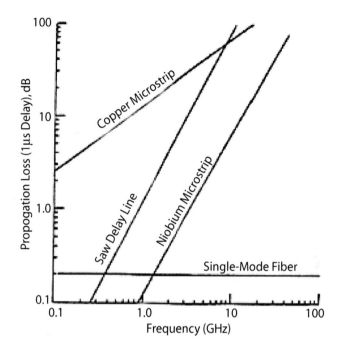

Figure 4–1 *Propagation loss in various delay lines.*

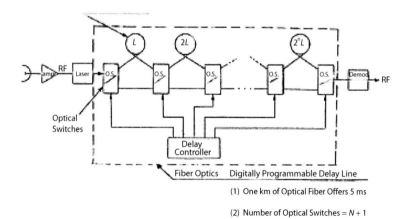

(1) One km of Optical Fiber Offers 5 ms

(2) Number of Optical Switches = $N + 1$

Figure 4–2 *Coherent RF memory using programmable FO delay line.*

threats with minimum cost and complexity. The DPODL architecture shown in Figure 4–3 directly converts the RF signal to an optical signal by direct modulation of a coherent gallium arsenide (GaAs) injection laser diode.

This particular delay-line design offers a very high quality RF memory device with maximum ECM effectiveness because it is capable of handling any type of frequency or phase-coded signal. A well-designed DPODL can provide a dynamic range greater than 50 dB, delay range from 0 to 655 microseconds in 20 ns increments, switching speed much less than one microsecond, triple transient response better than −35 dB, and delay variation not exceeding 1 ns over a wide temperature range. The dotted enclosure shown in Figure 4–2 contains SM optical fiber cable of appropriate length to meet the delay increment, optical switch (OS) modules integrated with directional couplers (DCs), and a digital delay controller circuit. This particular configuration will meet the requirements for genuine reproduction of frequency-coded waveforms with high accuracy. In addition, this delay-line architecture can maintain the spectral purity of noncoded waveforms, which is not possible with conventional RF recirculating-loop memory. In most cases, an optical amplifier may be required to compensate for the losses in optical switches, optical couplers, and optical fiber sections needed to meet delay requirements (see Figure 4–4).

This delay-line architecture is most attractive for a digital RF memory (DRFM) design configuration, which has a potential application in airborne ECM suits. DPODL design using FO technology offers compact size, reduced power consumption, and built-in immunity against electronic jamming and nuclear radiation, thereby making it best suited for the protection of costly, high-performance fighter and surveillance aircraft.

The overall delay with N sections each of a delay interval of 20 ns can be written as

$$T_0 = [(2^N)(T_i)] \qquad 4.1$$

where N is the number of bits required to meet a specific delay amount and T_i is the delay interval, which is assumed as 20 ns. Preliminary calculations indicate that a 15-bit design will require an optical fiber length of about 131 km to achieve a delay of 655.3 microseconds [1].

4.1.1.2 Description of Critical Components

OS modules integrated with directional couplers are the most critical elements of the DPODL. The delay-line design shown in Figure 4–5 involves double-pole-double-throw (DPDT) or 2×2 optical switches, 4-port directional couplers, and several sections of SMOF. Integration of optical switches and couplers on a single substrate (see Figure 4–5) will yield the most compact design with

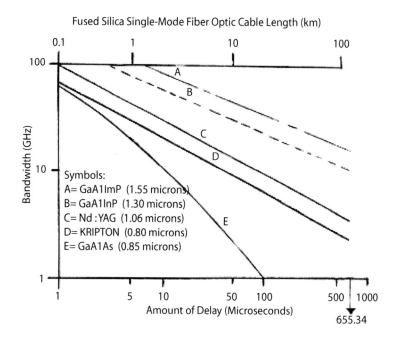

Figure 4-3 *Bandwidth and FO length as a function of optical source wavelength and delay.*

minimum cost and insertion loss. Each SMOF section is located between the OS and the optical coupler (OC). The output of the first OS goes through the first SMOF section with a predetermined delay that finally terminates into the first OC. When the OS is on the "ON" position, the optical signal enters the other input port of the OC. This process continues until the final signal with the desired delay appears at the output of the last OS. Each delay-line section will affect the overall system bandwidth, dynamic range, spurious levels, triple-transit response, and delay variation over the RF bandwidth and temperature range. Performance requirements for the digital control unit consisting of a digital-to-analog converter (DAC) and MUX bus are dependent on the optical switch modules.

4.1.1.3 Performance Requirements for the Optical Coupler (OC)

This device exhibits many characteristics, which makes it a highly efficient power divider. The OC has a typical loss of 0.5 dB and directivity greater than 50 dB. The device is practically independent of the input polarization. and the polarization loss in an SMOF is less than 0.1%. The optical T-coupler is an equivalent of the standard microwave coupler with minimum loss. The unclad

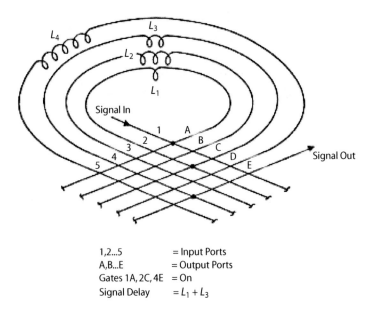

Figure 4–4 *FO switch array configuration involving fiber delay sections to form programmable delay line.*

optical fibers must be joined using transparent cement with a refractive index matching closely to that of the fiber used. The most efficient coupling is achieved when the refractive index of the cement is equal to that of the optical fiber. The coupling factor (K) applies for either direction of the optical-power transfer. Excessive coupling losses can be expected, resulting from discontinuities within the coupler formed by the mechanical misalignment.

Coupling losses in a coupler are strictly dependent on the NA of the fiber, operating wavelength, and discontinuities (if any) within the fiber. The coupling loss can be calculated on the assumption that all the light reaching the end face of the cement volume is completely lost. The loss contributions from several patterns of two, three, and four cemented fibers must be considered. Each loss factor is proportional to the ratio of the area occupied by the fiber cross-section to the total area of the fibers involved.

4.1.1.4 Requirements for Optical Switches

The OS provides the switching function from port 1 to port 2 with minimum signal loss. The use of an OS array in conjunction with optical fiber delay sections to form a programmable delay line is illustrated in Figure 4–4. The input and output ports are clearly marked, and the amount of delay at each fiber section

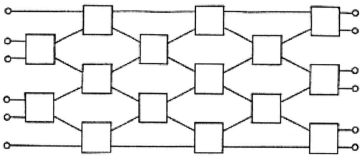

Integration of 2x2 Single-Mode Optical Switches on a Single Substrate

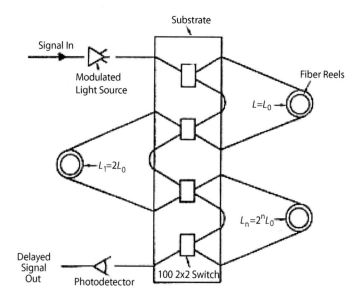

Figure 4–5 *Critical elements of a programmable optical delay using light source, optical switched integrated on a single substrate, fiber reels, and photo detector.*

is accurately specified. As stated earlier, the OSes and couplers can be integrated on a single substrate (see Figure 4–5) if minimum cost and size are the critical requirements. The state of each of the OSes determines which of the interconnected SMOFs transmits the signal to the next switch downstream. A fused silica SMOF is best suited for the coherent DPODL application.

The author has investigated several types of OSes, including thermo-optic, electro-optic and acousto-optic switches, but none of them offers performance close to FO-based switches. MEMS OSes have been considered; nevertheless, these switches are not suited for this particular application because of poor switching speed and high insertion loss.

4.1.2 Impact on Delay-Line Performance from Various Sources

This subsection will summarize the impact on delay-line performance from optical component parameters and operating environments such as temperature, humidity, and pressure. Component parameters include spacing between the fiber taps, fiber surface conditions, and material impurities.

4.1.2.1 Impact Due to Temperature and Tap Spacing

Delay-line performance remains unaffected due to temperature fluctuations so long as the operating temperature does not exceed 200°C. The performance of an FO-based delay line is only affected by the gross effects that change the optical length by an amount on the order of the modulation envelope of the RF signal. Preliminary calculations performed by the author reveal that a change of 150°C in temperature, which is highly unlikely to occur, would at most cause a delay variation of 0.1% in the delay interval. This amounts to a delay variation of less than 0.03% over a temperature range from 0 to +50°C, which will cause a delay variation of less than 0.6 ns in a delay interval of 20 ns.

Time-delay variation is also dependent on the spacing between the fiber taps. The above calculations further indicate that the time-delay variation between two closely spaced 20-ns taps is less than 5 ps if the spacing between the fiber taps is within 1 mm. Delay variations as a function of numerical aperture, spacing tolerance, and source wavelength have not been computed to avoid excessive computational efforts.

4.1.2.2 Impact on Delay-Line Performance from Fiber Parameters

Time-delay variations are also dependent on optical fiber parameters such as NA, core diameter, and refractive index as a function of optical wavelength. Optical fibers made from fused silica are best suited for use in coherent optical delay lines irrespective of delay amount. SMOFs made from fused silica provide large dynamic range, low transmission loss, and minimum modal dispersion because all of the optical power coupled into an SMOF propagates in a single transmission mode. An SMOF with a core diameter of 4 microns exhibits insertion loss of 3 dB at 0.83 microns, 1 dB at 1.03 microns, and 0.35 dB at 1.33 microns in wavelength. In the absence of modal dispersion, high frequency signals can be transmitted with minimum dispersion loss over a wide band.

Modal dispersion generally occurs in an MM fiber and results in severe signal degradation leading to a pulse-broadening effect. This effect is proportional to the square root of the dielectric constant of the core material. In general, core and cladding dimensions and the refractive-index profile of the optical fiber determine the modal dispersion impact on the delay-line output pulse.

4.1.2.3 Modulation Bandwidth of Optical Fibers

The modulation bandwidth of an SMOF is limited by the dependence of the fiber refractive index on the carrier wavelength. The 3 dB modulation bandwidth of an SMOF below a carrier wavelength of 1.06 microns is inversely proportional to root mean square (RMS) half-width of the optical source, to the differential dispersion per unit length, and to the optical fiber length. Studies further indicate that in optical fibers of short length [2], material dispersion is not significant. Furthermore, the material dispersion in short optical fibers made from silica disappears at an operating wavelength of 1.27 microns. The modulation bandwidth is also limited by group delay variations contributed by other optical components in the delay-line structure and by polarization dispersion due to two polarization modes supported by the optical fiber. A modulation bandwidth of 100 GHz is possible in a fused-silica core fiber with an optical-source line width ($\Delta\lambda$) of 1 Å and beam broadening of 10 ps. If the beam broadening increases to 64 ps, the bandwidth is reduced to 16 GHz.

4.1.2.4 Impact of Operating Wavelength on Chromatic Dispersion

Material dispersion, also known as chromatic dispersion [2], is a critical performance requirement in coherent delay-line design. The material dispersion (D_{mat}) can be written as

$$D_{mat} = [(\lambda/c)(d^2n/d\lambda^2)] \text{ ps/nm-km} \qquad 4.2$$

where λ is the optical-source wavelength, c is the velocity of the light, and n is the refractive index of the core material.

Material dispersion is dependent on the source spectral bandwidth ($\Delta\lambda$) and is present in both the SM and MM optical fibers. The material dispersion is directly proportional to source wavelength and is a second derivative of the refractive index (n) with respect to source wavelength (λ). The material dispersion is extremely low at a source wavelength of 1.33 microns. The spectral bandwidth must be kept as narrow as possible to minimize the material or chromatic dispersion.

4.1.2.5 Impact of Source Wavelength on Delay-Line Performance

With long optical fibers, it is desirable to operate at wavelengths where the differential group delay ($dT_g/d\lambda$) is close to zero. The symbol T_g indicates the group delay per unit length and λ indicates the source wavelength. Optical sources at longer wavelengths are preferred to minimize the differential group delay. Based on the author's design and development experience, the bandwidth-distance product for a fused-silica-based SMOF is greater than 200 GHz-km. However, a time-bandwidth product greater than 10 is required to achieve a delay of 655.3 microseconds over an RF bandwidth of 16 GHz, as illustrated in Figure 4–2.

4.1.2.6 Overall Delay-Line Performance Summary

Trade-off studies must be performed in terms of bandwidth, delay amount, source wavelength, and mechanical tolerances to meet the overall system performance requirements. Various loss mechanisms due to mechanical misalignment, the splicing process, the critical bend radius, and the optical cement thickness can affect the dynamic range and bandwidth of the delay line. Frequency dispersion, polarization dispersion, and multiple scattering can also affect the overall-delay line performance. Raman scattering is independent of the RF frequency, whereas the quantum noise is linearly proportional to signal bandwidth. This indicates the optical dynamic range of an FO-based delay line decreases linearly with increasing signal level for a given operating wavelength and given amount of delay.

4.2 FO-Based Michelson Array

Polarization-induced signal fading is not acceptable in covert communications systems used by battlefield commanders and/or personnel and intelligence agencies. Therefore, a device capable of eliminating the polarization-induced signal fading is critical in covert communications systems. An FO-based Michelson array (FOMA) using a Faraday-rotator/mirror device offers complete elimination of polarization-induced signal fading. This passive elimination technique uses a polarization-selective element in the input to provide adequate isolation of the source from the reflective array.

The Michelson's interferometer (MI) technique uses a 45 degree Faraday rotator followed by a plane mirror. The use of a nonreciprocal rotation element in the MI configuration provides passive stabilization against polarization fading due to environmental birefringence perturbations.

4.2.1 Description of All-Fiber Version of MI

This subsection describes an all-fiber version of an MI using Faraday-rotator/mirror (FRM) packaged devices for the birefringence compensation (BC). This version of interferometer uses a polarization-selective element in the optical fiber to provide high isolation of the optical source from a very strong return optical signal. The MI block diagram shown in Figure 4–6 illustrates a system configuration that has demonstrated a polarization-independent MI device with pigtailed FRMs as the reflector elements.

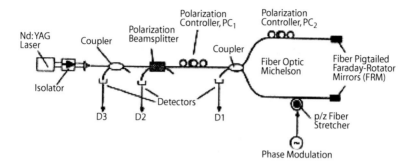

Figure 4–6 *Block diagram of polarization-independent, FO MI array.*

This system configuration (see Figure 4–6) involves a linear in-line optical isolator, optical couplers, a detector, a polarization-beam splitter, a polarization-fiber stretcher, an FO-based Michelson Array (MA), and two FRM elements. Each packaged device consists of a Faraday rotating element and a dielectrically coated mirror. Overall reflector efficiency can be achieved in excess of 75% with minimum cost and complexity. A 1.3 micron laser beam enters the system via a 3 dB optical coupler and polarization-beam splitter. The pigtailed FMS elements are included in both arms. A sinusoidal 360-degree phase shift at 5 kHz is applied to the piezo-detecting fiber stretcher. Fringe visibility can be monitored by detectors located at various positions shown in Figure 4–6. Isolation between the source and return signal is dependent on the state of polarization (SOP) of the splitter. Because of various reflecting components, isolation better than 25 dB is not possible [2]. However, additional isolation can be achieved by inserting a fiber polarizer between the optical coupler and beam splitter.

4.2.1.1 Description of a Multielement System Using Michelson Array

It is possible to overcome polarization-induced fading using active control of the input state of polarization. Full optimization of the fringe visibilities of an array of sensors using polarization tracking is not feasible because the input SOP required to optimize the visibility of one sensor will not correspond to that required for another sensor in a four-sensor, time-division-multiple-access (TDMA) system with FRMs acting as the reflectors (see Figure 4–6). An acousto-optic modulator is used to obtain a pair of pulses of duration T with a separation of T_s. FO couplers of varying splitting ratios and FRMs with matching reflection efficiencies are selected to achieve return pulses of nearly equal intensity through the polarization beam splitter to the optical detectors. A frequency modulation is applied to the laser to generate an interference signal for the four sensors involved. A fringe visibility greater than 0.95 simultaneously for all the sensors under birefringence perturbations is induced manually in the fiber leads.

This system architecture has demonstrated a concept of passive control of polarization-induced fading in an array consisting of multiple sensors. No significant polarization-induced phase variations are visible in a sensor locked at maximum visibility. This scheme of simultaneously optimizing the fringe visibilities of all the sensors in a four-sensor MA is completely passive and is best suited for applications where isolation better than 25 dB between the source and the return signal is desired.

4.3 Tunable Dispersion Compensation

Stringent dispersion performance requirements are essential for communications systems used by military and intelligent agencies to preserve maximum security and covertness. Tunable dispersion compensators must be deployed in communications systems or transmission systems carrying high data rates from 40 to 100 Gbits/sec to meet a low bit error rate if security and covertness are the principal requirements. Tunable dispersion compensators (TDCs) employing two quadratically chirped FBGs can solve the high-order dispersion problems in long-haul communications systems.

TDC devices can be fabricated using various emerging technologies, such as a virtually imaged phased array, cascaded Mach-Zehnder planar lightwave circuit, reflective etalon with MEMS-based variable reflector, tunable ring resonator, and FBG. Most of these device technologies exhibit higher-order dispersion due either to defective design or to fabrication problems. However, a TDC design employing two quadratically chirped FBGs does not suffer from higher-order dispersion problems. This particular TDC device design has a linear dispersion characteristic and a simple tuning mechanism [3].

4.3.1 TDC Configuration using Two Quadratically-Chirped FBG Devices

The TDC configuration using two quadratically-chirped FBG (QCFBG) devices is shown in Figure 4–7. This particular design uses two FBG devices coupled with a four-port optical circulator, and each FBG has a quadratic group-delay reflection characteristic. The first FBG device has positive linear and quadratic delay terms, whereas the second FBG device has negative terms with the same magnitudes. Both FBGs are attached to an independent piezoelectric (PZT) actuator, which linearly strains the optical fibers resulting in a wavelength shift in their group-delay characteristics. The dispersion is determined by their wavelength separation. When both gratings are strained to their midrange, zero dispersion occurs with spectral coinciding. Positive dispersion is obtained by simultaneously reducing the tension in the FBG device A and increasing the tension in the FBG device B as illustrated in Figure 4–7. Similarly, negative dispersion can be achieved in the other direction.

The quadratic group delay is obtained with a quadratic-chirp profile based on the assumption that discrete wavelengths are reflected from single points on the grating surface.

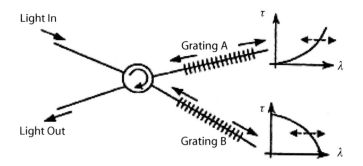

Figure 4–7 *Dispersion-compensator design configuration using chirped filter gratins.*

The change in the Bragg wavelength as a function of distance along the grating axis can be expressed as

$$[\Delta l_b] = [-D_0/\varepsilon] + (1/\varepsilon)[D_0^2 + 4n\varepsilon z/c] \qquad 4.3$$

where D_0 is the linear dispersion at $z = 0$ (ps/nm), n is the refractive index of the optical material, z is the distance along the grating axis (mm), ε is the dispersion slope $(dD_0/d\lambda)$ in ns/nm^2 and c is the velocity of light (mm/sec).

The choice of dispersion slope requires a trade-off in terms of critical parameters, including optical signal bandwidth (B_o) and the length constraint of the gratings. Variations in Bragg wavelength is a function of axial distance (z) and the dispersion to be tolerated, as illustrated in Figure 4–8. The curves shown in Figure 4–9 indicate that the Bragg wavelength increases with an increase of axial distance but decreases with the increase in linear dispersion (D_0) as a function of axial distance (z). Furthermore, increasing the dispersion slope will result in a greater change in linear dispersion for a given change in wavelength separation of the gratings and in a reduction in reflection bandwidth (B_r) of each grating, such that the maximum wavelength separation available is reduced [3].

The dispersion tuning range of the compensator can be written as

$$T_R = [2\varepsilon(B_r - B_0)] \qquad 4.4$$

Computed and measured values of the tuning range as a function of dispersion slope for a single grating are shown in Figure 4–8.

The grating length (L_g) is dependent on the size constraint of a mechanical actuator. Based on preliminary calculations, a grating length of 100 mm is adequate, including the apodization regions representing the major portion of the length. Using the optimum dispersion slope, an optical bandwidth of 0.5 nm is just right for 40 Gbit/sec operations. The tuning-range plots as a function of dispersion slope (ε) and linear dispersion (D_0) are shown in Figure 4–8. It is desirable to make the linear dispersion as small as possible to achieve gratings with low dispersion and high chirp, which are critical requirements for the optimum performance of a dispersion compensator. Preliminary computations indicate that a dispersion of 200 ps/nm will yield optimum performance.

4.3.1.1 Acceptable Dispersion Limits in Covert Communications Systems

Dispersion can limit the performance of a WDM telecommunications system or dense-WDM communications systems. Optimum system performance is possible with low chromatic dispersion with a typical value of a few picoseconds per nanometer per kilometer. With a very low dispersion, the optical signals may stay in phase over long distances. A little dispersion spreads the signals out, thereby reducing their interaction and damping four-wave mixing. Too much dispersion limits transmission distance and requires dispersion compensation devices. A further complication comes from the slope of the dispersion curve over broad bandwidth of a WDM or dense-WDM system. Dispersion in an SM optical fiber varies with the operating wavelength. For an SM fiber, the dispersion slope is about 0.08 ps/nm^2-km at a wavelength of 1550 nm. If the optical fiber is to carry signals over a 100 nm optical bandwidth, the chromatic dispersion can vary as much as 8 ps/nm-km. Thus, the lower-dispersion wavelengths can travel five times further than the higher-dispersion wavelengths before they require dispersion compensation. In addition, optical fibers with lower dispersion slopes are best suited for wider optical bandwidths.

4.3.1.2 Impact of Temperature on Bragg Wavelength

The temperature dependence of the Bragg wavelength (λ_b) in an FBG is typically about 1 nm per 100°C, which is too large for a DWDM system. The temperature dependence of the Bragg wavelength for an uncompensated FBG can be given by

$$[d\lambda_b/dt] = (2g_p)[dn/dT + n\alpha_f + ds_{long}/dT] \qquad 4.5$$

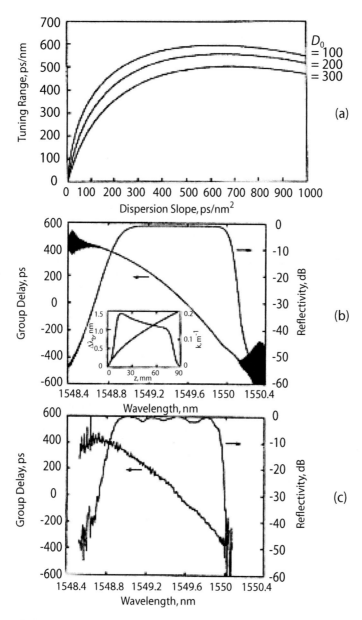

Figure 4–8 *Compensator performance: (a) tuning range versus slope, (b) group delay versus wavelength, and (c) reflectivity versus wavelength.*

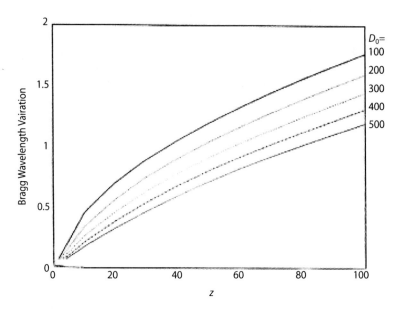

Figure 4–9 *Bragg wavelength variation as a function of position (z) along the grating axis for various values of linear dispersion (D_0).*

where g_p is the grating period, n is the refractive index, T is the operating temperature, t is the coefficient of thermal expansion, and s_{long} is the longitudinal strain. To produce a longitudinal strain, one requires an aluminum block on each end of a silicon glass tube containing the FBG. The coefficient of thermal expansion for the aluminum is $25 \times 10^{-6}/°C$ and for the silica is $0.55 \times 10^{-6}/°C$. The differences produce a strain with negative temperature dependence of the Fiber BRAGG Grating (FBG) assembly. Adjusting the ratio of the aluminum-block length to the silica-tube length controls the temperature dependence of the Bragg wavelength. Temperature dependence of less than 0.5 per million/°C is possible from a temperature-compensated FBG over a temperature range of $-40°C$ to $+80°C$. This corresponds to a temperature dependence of less than 0.05 nm/100°C. This is considered ideal for a covert DWDM communication system.

4.3.1.3 Insertion Loss, Reflectivity, and Group Delay in FBG Devices

The reflectivity parameter of an FBG device is dependent on its chirp and operating wavelength. Sometimes it is necessary to adjust the coupling coefficient to compensate for the adverse effects arising from the reflector surface. The coupling-coefficient profile is multiplied by an apodization function [4] to minimize the group-delay ripple. The coupling-coefficient profile parameter for a single FBG device as a function of axial location (z) is shown in Figure 4–8.

Group-delay and reflectivity characteristics as a function of wavelength are also shown in Figure 4–8. The simulated reflectivity with flat spectral response is possible over the 1540.8-to-1549.8 nm range without any apodization induced ripple. However, in actual practice, some ripple in the group delay and reflectivity curves can occur, as shown in Figure 4–8, due to parametric tolerances in two FBG devices. Group delay for the entire TDC module as a function of operating wavelength will be quite different from that of the FBG devices, as illustrated in Figure 4–10. The insertion loss in an FBG device is constant with tuning, unlike nonlinearity-strained or temperature-tuned FBG devices.

4.3.1.4 Q-Penalty With and Without Tunable Dispersion Compensator

The overall system performance can be judged by looking at the Q-penalty curves and eye diagrams as a function of dispersion with (dotted curve) and without (solid-circle curve) integration of a TDC in the system. The system has a maximum Q of 15 without TDC and operates over a 100 ps/ns dispersion range with a Q-penalty of 2 dB, as shown in Figure 4–10. With integration of a TDC, the system can operate over a net dispersion range of 600 ps/nm with a maximum penalty of 1.7 dB. It is interesting to mention that a temperature-tuned FBG device has a tuning range of 170 ps/nm in a 40 Gbit/sec system. In a TDC module using two quadratically chirped FBG devices, the dispersion penalty is increased to 600 ps/nm in the same 40-Gbit/sec system. This indicates that twin fiber chirp grating configuration yields much better performance over other dispersion-compensating techniques.

4.4 Optical Ring-Resonator Gyros (ORRGs)

Several types of FO gyros have been developed for applications in industrial, commercial, military, and space sensors. However, a passive ORRG, which uses a finesse fiber ring-resonator concept, offers better performance with a much shorter optical fiber loop than an interferometer-type FO gyro. This particular gyro design configuration is best suited for applications in fighter aircraft, missiles, and space sensors, where weight and size are the most critical requirements. However, bias induced in an ORRG because of the optical Kerr effect contributes to a dominant noise source. This bias is proportional to the difference in intensity between CW and coded-CW (CCW) light waves present in a resonator. Even a small imbalance of 0.01% between the two light waves produces a bias larger than the shot noise by two orders of magnitude [5].

Kerr-effect-induced bias can be effectively eliminated using proven techniques. In one method, the bias is monitored by modulating the light-source intensity and fed back into one light wave-intensity signal traveling in the resonator to make it zero. Note the Kerr-effect-induced drift can slightly degrade the

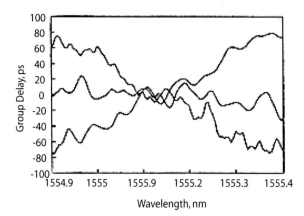

(a) Group Delay as a Function of Wavelength for the Entire Module

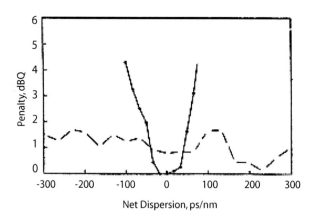

(b) Q-Penalty for the System With (Solid Line) and Without Tunable Dispersion Compensator

Figure 4–10 *(a) Group delay and (b) Q-penalty for the system with and without TDC.*

gyro performance. But this induced drift can easily be reduced by adjusting the gyro parameters.

Once the induced bias and induced drift are eliminated or significantly reduced, the ORRG has potential applications in unmanned air vehicles (UAVs), missiles, high-performance fighter aircraft, reconnaissance platforms, and space sensors where precision gyro performance is critical for meeting mission requirements.

4.4.1 Techniques to Reduce Kerr-Effect-Induced Bias Error

Limited techniques to reduce Kerr-effect-induced bias errors are available. However, an experimental technique illustrated in Figure 4–11 offers the most effective method for reduction of Kerr-effect-induced bias errors. The experimental setup offers a cost-effective approach to reduce such bias errors. This particular method uses a ring YAG laser ($\lambda = 1.3$ microns) with a spectral line width ($\Delta\lambda$) of about 300 kHz including the random jitter. The optical resonator consists of a PM optical fiber and coupler. The resonator length and diameter are 1200 cm and 10 cm, respectively, and its finesse is close to 100. A resonator with a 90-degree polarization-axis rotation is required to suppress the polarization-fluctuation-induced bias error. The piezoelectric transducer (PZT) shown in Figure 4–11 is modulated by a 7 kHz sinusoidal wave to catch the resonant point by the lock-in amplifiers (LIAs). The output of the detector D_2 is fed back into the PZT through LIA#2 to cancel the environmental fluctuations due to temperature and pressure. The laser frequency is controlled through a digital feed scheme. Closed-loop operation is necessary to remove the scale-factor change, if any. The rotation signal is obtained as the frequency difference between AOM1 and AOM2, as shown in Figure 4–11. Noise interference between the signal wave and backscattering in the optical resonator is reduced by a binary phase shift keying (BPSK) modulation. This eliminates the carrier components of the CCW wave.

The light-wave intensity into the resonator is slightly modulated with a sinusoidal modulation frequency parameter (k) using the intensity modulator unit 1 (IM1). The modulation frequency is much lower than the resonator point detection q. The differential output P_3 between LIA1 and LIA2 contains the frequency component k that is produced by the Kerr effect. The Kerr-induced bias is synchronously detected with the LIA3 module, and the output port P_4 of LIA3 is fed back into the intensity-modulator module 2 (IM2) to make the intensity difference zero. The YAG laser has an intensity-modulation function, which is used in the test phase in place of IM1. The intensity modulation itself by IM1 can cause an additional k component, which can affect the resonator output. The resonator output $g(t)$ is a function of resonant frequency (f_0), frequency deviation of the modulation frequency (f_w), Kerr-effect-induced bias, offset of the resonant point, and the phase term of the intensity-modulation function. The demodulation output is nothing but the derivative of the output with respect to frequency.

Mathematical analysis indicates that the intensity-modulation component is very small compared with the Kerr-effect-induced k components and hence can be neglected. The k component related to the Kerr effect at LIA3 can be detected. The output of the detectors D1 and D2 is divided by the monitor output of the YAG laser. Finally, the Kerr-effect-induced bias is compensated for by the intensity change.

The time constant for the gyro output varies from 25 to 30 seconds. The initial CW light wave intensity in the resonant structure is about 55 microwatts,

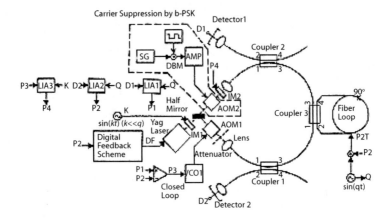

Figure 4–11 *Experimental setup to evaluate the reduction method of Kerr-effect-induced bias error.*

which can be reduced to 42 microwatts using a variable attenuator at the test points indicated by the arrows shown in Figure 4–11. The Kerr-effect-induced drift is about 10^{-5} radian/second and is limited by the Kerr effect. Further reduction in Kerr-effect-induced drift is possible by adjusting the critical parameters of the system as mentioned above.

4.4.2 Dispersion Compensation Using MEMS-Based Mirror Technology

A fixed dispersion compensator, such as a chirped FBG or dispersion compensating fiber, can eliminate or reduce the steady-state dispersion in the system but cannot compensate for the dynamic dispersion effects [6]. Note chromatic dispersion in optical fibers distorts a signal by stretching the duration of the pulses, thereby causing them to overlap, as well as increasing the bit error rate (BER) of an optical system such as a WDM or DWDM communications system. To overcome these problems, one needs a mechanism capable of compensating the dynamic dispersion.

4.4.2.1 Dynamic Dispersion Compensation System Using MEMS Technology

This particular compensating system uses mechanically stressed FBG devices that are capable of providing tunable compensation of individual channels in a WDM communications system. In brief, this MEMS-technology-based system uses micromirrors and provides tunable dispersion compensation across the spectrum of channels of a WDM system. This compensation technique essentially offers dynamic dispersion compensation over all channels in a single package.

In a multichannel, tunable dispersion compensator that is capable of providing dynamic dispersion compensation, the input optical beam passes through a three-port optical circulator and is focused in free space onto a bulk diffraction grating (BDG), as illustrated in Figure 4–12. This grating separates the WDM channels and spreads out the frequency components present in each channel. The frequency components are then incident on the MEMS mirrors.

The electrostatic force produced by the actuator distorts the surface of the flexible mirror elements, thereby imposing a different phase shift on each reflected beam. A different phase-shift correction is applied to the frequency components, depending on the voltage applied to the electrostatic actuator. The change in the curvature of the mirror surface is a small fraction of a wavelength, which causes negligible angular displacement of the reflected radiation. The reflecting frequency components are combined at the diffraction grating. The optical beam is coupled back into the optical fiber and emerges from the third port of the circulator. The optical layout of the compensating mechanism can be designed to provide greater separation between the MEMS mirror, which will provide channel spacing as small as 1 nm. The dispersion can be controlled for each of the three channels and varies from 4 to 16 ps/nm. This particular compensator architecture has demonstrated a flat, zero-voltage insertion loss at 10 dB across each channel that slightly increases when full voltage of 100 volts is applied. An improved system layout and precision optical alignment can reduce this insertion loss well below 5 dB.

4.5 All-Fiber, Q-Switched (AFQS) Laser

An AFQS laser, as shown in Figure 4–13, incorporates a ring resonator and is capable of achieving a high-power, single-frequency laser while eliminating the standing waves. This can burn holes in optical resonators. FO lasers with moderate power levels are commercially available and promise to be more powerful as the technology matures in the near future. Initially, the average power obtained was 2 mW at a wavelength of 1557 nm, whereas the ring lasers demonstrated higher power levels at the same wavelength. In the AFQS laser in Figure 4–13, the ring resonator from a loop is on the left side, and the resonator including a Q-switch and a Mach-Zehnder interferometer is on the right side. The PZT is located between the two optical couplers. The pulse duration is typically 0.5 microseconds for the AFQS laser, and the Q-switched pulse repetition frequency (PRF) can be as high as 800 Hz. The spectral line width of the pulse can be as low as 25 MHz.

High-energy AFQS lasers have potential applications in LIDARs, laser range finders, and distributed sensors. The AFQS lasers outperform the crystal Q-switched lasers that exhibit higher insertion and coupling losses. A single-frequency laser is best suited for specific military applications. Studies further indicate that the constructive or destructive interference within the interferometer

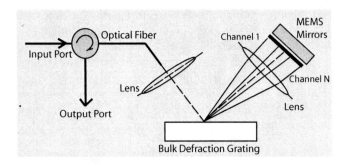

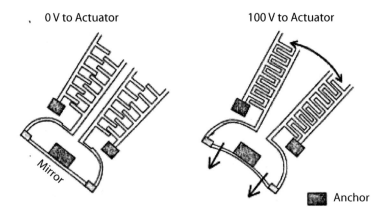

Figure 4–12 *Distortion of the flexible mirror surfaces due to application of electrostatic force exerted by the actuator.*

is dependent on the phase change introduced by the PZT. In the case of destructive interference, very little light is returned to the ring resonator, and the Q-switch eliminates the oscillations in the laser. However, in the case of constructive interference, light energy returns to the resonator, and the ring laser will oscillate at high power levels.

4.5.1 Critical Design Aspects and Parameters

A small linear resonator FO laser located between the two distributed Bragg reflectors (DBRs) is pumped with a solid-state laser diode emitting at 980 nm. The effective length of the resonator is 30 mm, and the laser oscillates in a single

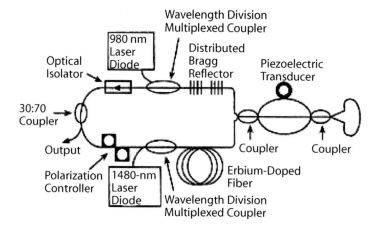

Figure 4–13 *AFQS laser incorporating a ring resonator to yield high-power, single-frequency oscillation.*

longitudinal mode. The narrow-bandwidth photons generated by the solid-state laser act as seeds to force the ring laser into a single-frequency operation. The ring laser that is pumped with a 1480 nm diode laser oscillates weakly in a single mode, even when the Q-switch is on "OFF" position. When the switch is in the "ON" position, the single mode quickly builds the amplitude until it saturates the gain. Several centimeters of optical fibers in the seed laser are doped with erbium at a concentration of 700 part per million, whereas the 400 centimeter long fibers are doped at 280 parts per million. The optical isolator shown in Figure 4–13 ensures single-direction oscillation around the ring, while the polarization controller guarantees the correct polarization. The optical output signal is obtained with the 30% optical coupler shown in Figure 4–13.

4.6 Solid-State Beacon Laser Illuminator (SSBLI)

The optical beacon laser illuminator is the most critical element of the beam control/fire control system in a high-power airborne laser capable of destroying hostile airborne ballistic missiles in their boost phase. In other words, the SSBLI plays a key role in antimissile defense systems by manipulating the phase conjugation on the high-power chemical oxygen iodine laser (COIL). The destruction of the incoming missile is accomplished by focusing the "killer beam" of the COIL laser on the target. The airborne high-power laser output power requirement depends on the incoming ballistic missile's range, operating altitude, and atmospheric conditions. The term "killer beam" is used because the laser beam deployed offers the most effective process of cutting, melting, and surface decontamination in the shortest time possible. Both the carbon dioxide and COIL lasers are capable of producing high output power levels [7]. A significant increase in laser output and conversion efficiency is possible by injecting atomic rather than molecular iodine into a COIL system. The COIL laser offers both the high-power CW and pulsed levels of near-diffraction-limited quality and a near-perfect circular optical beam with high conversion efficiency. Scientists at the University of Illinois [8] have successfully demonstrated the cutting of a half-inch-thick steel plate with a 7.36 kW COIL laser. The high-power, solid-state laser illuminator can be deployed to determine the wave front error caused by the atmospheric turbulence between the hostile target or the incoming ballistic missile.

4.6.1 Description of Critical Elements

The design of the airborne laser illuminator depends on the tactical mission requirements. The critical elements of the airborne laser illuminator include the high-energy laser modules, solid-state diode-pumped Nd:YAG slab laser (also called a master oscillator) stimulated Brillouin scattering cell, beam control device, and optical power amplifiers [8]. Each of the two line-replaceable units (LRUs) features an optical oscillator producing a 1064 nm pulse that is injected into three optical amplifiers. After amplification, the illumination beam is focused into a stimulated scattering cell, which uses sound waves to correct the induced aberration. Deployment of nonlinear optics, compact packaging, efficient cooling, and precision optical alignment is essential if small size, light weight, and reliability are the principal requirements for the airborne laser illuminator.

4.6.2 Impact of Atmospheric Parameters on Wavefront Errors

As mentioned earlier, the laser illuminator emits optical energy to determine the wavefront created by the turbulence effects in the atmospheric region between the high-power chemical laser and the incoming ballistic missile. The turbulence is caused by atmospheric aerosols, which are influenced by the variations in temperature, pressure, air density, and molecular weight as a function of altitude and operating wavelength. In other words, the intensity of turbulence is dependent on the atmospheric parameters as a function of operating altitude and wavelength. The reflected optical signal from the hostile incoming missile passes through a series of wavefront monitoring sensors abroad the surveillance aircraft. These monitoring sensors will then direct the deformable mirrors to compensate for the optical-signal distortion created by the turbulence effects in various atmospheric layers. The lower atmospheric layer known as the troposphere varies from 7 to 10 miles above the earth's surface, whereas the upper atmospheric layer known as the stratosphere varies from 10 to 30 miles above the earth's surface. Aerosols in the upper atmosphere represent only a small fraction of the total contents of the aerosols. However, lasers operating in urban areas or in the lower atmosphere will experience high levels of aerosol attenuation and scattering because of impurities present in the region. Due to the presence of aerosols, an optical signal will experience extinction, absorption, diffraction, and scattering. The intensity of the turbulence is dependent on the atmospheric characteristics and urban and rural environments.

Atmospheric pressure, density, and molecular weight all decrease with an increase in altitude. Studies performed by the author [9] indicate that absorption and diffraction are negligible at altitudes of 30 km or more, but volumetric scattering, multiple scattering, and molecular optical depth will have serious effects at altitudes from 5 to 20 km. Typical values of volumetric scattering $K(\lambda, h)$ and molecular optical depth $Q(\lambda, h)$ as a function of wavelength (λ) and altitude (h) are shown in Table 4–2.

The tabulated data indicate that volumetric scattering will have significant impact on laser signals in the boost phase due to the high intensity of turbulence and the large size of aerosols present in the lower atmospheric layer.

Effects from volumetric scattering are more pronounced when the particle size is compatible to the operating wavelength. Volumetric scattering is inversely proportional to the fourth power of the particle size. Both multiple scattering and volumetric scattering can significantly bend the laser beam because both these quantities are dependent on the particle size and complex refractive index of the aerosol as a function of wavelength. The particle size in the lower atmosphere varies from 0.1 to 10 microns depending on the atmospheric conditions at a given altitude. Sometimes, the decrease in refractive index with sensor altitude or height may be so great that the laser beam is bent downwards. In contrast, under certain atmospheric conditions, the refractive index may increase with height, causing the optical signals to bend upwards. In summary, the refractive index is a function of temperature and wavelength and decreases when sensor height increases. The variation in the refractive index (n) as a function of temperature (T) varies from zero at sea level to 0.42×10^{-6} at 10 km height to 0.03×10^{-6} at 25 km height.

Table 4–2 *Typical values of volumetric scattering and molecular optical depth.*

Altitude (km)	Volumetric Scattering λ (nm)			Molecular Optical Depth λ (nm)		
	800	900	1000	800	900	1000
5	0.0015	0.0010	0.0005	0.011	0.005	0.004
10	0.0009	0.0005	0.0003	0.006	0.004	0.002
15	0.0002	0.0001	0.0001	0.001	0.001	0.001

4.7 FO Ring Laser (FORL) Gyros

FORLs use high-performance optical fibers to maintain optimum gyro accuracy and stability under severe operating environments. These severe operating environments include high magnitudes of shock, extreme temperatures, and nonlinear, multifrequency vibrations. Optical fibers for use in FORL gyros must meet stringent performance requirements to ensure high reliability, accuracy, and stability under severe operating conditions. These gyros offer excellent accuracy in measurements of linear and angular displacements under variable roll, pitch, and yaw conditions, which is difficult to achieve from other conventional devices. FORL gyros have potential applications in missile guidance systems, airborne reconnaissance and surveillance sensors, and other sophisticated military weapon systems.

During the launch phase, the missile undergoes severe shock and vibration. It is critical to provide high missile-guidance accuracy during the launch phase, cruise phase, and terminal phase to accomplish the mission objectives with high reliability. Performance of the missile guidance system during launch and terminal phases is critical. The gyros deployed must provide second-by-second, reliable and accurate information on the range to target, missile velocity, and angular positions during various phases of the missile flight. Optimum gyro performance requires elimination or suppression of the intensity noise generated by mode beating.

4.7.1 Beat-Noise Elimination Techniques

An effective beat-noise elimination technique is required to achieve reliable and optimum gyro performance. Trade-off studies performed on various elimination techniques leads the author to believe that a Lyot filter [10] provides effective suppression of the high-intensity noise generated by the mode beating. In many optical and telecommunications systems, EDFAs are deployed to meet system gain requirements. However, the most serious barrier to their use is the intensity noise that is generated when the laser longitudinal mode beats against the amplified spontaneous emission in the suppressed longitudinal modes. This noise can be suppressed by inserting a semiconductor optical amplifier, shown in Figure 4–14, into the ring resonator. The semiconductor optical amplifier (SOA) acts as a high-pass filter that effectively blocks the lower-frequency beat noise.

Since a FORL's resonator has a typical length of several meters, the beat frequency between the adjacent longitudinal modes is relatively low, meaning on the order of 10 to 12 MHz. The phenomenon of self-gain modulation in an SOA prevents it from amplifying these low-frequency components and their harmonics. Self-gain modulation occurs because the low-frequency intensity fluctuation on the incoming light depletes the carrier density in the SOA, causing its gain to saturate. This carrier depletion causes gain saturation at frequencies into the gigahertz (GHz)

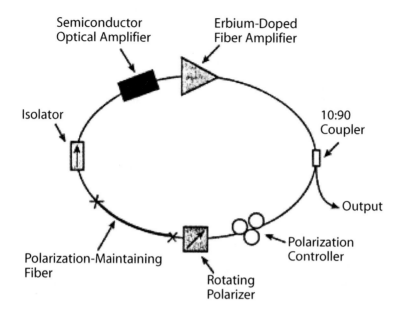

Figure 4–14 *Deployment of an SOA to provide noise suppression in an FORL.*

range. Limited carrier lifetime prevents self-gain modulation from occurring at higher frequencies. The polarization controller, rotating polarizer, and PM fiber shown in Figure 4–14 comprise a Lyot Filter. This forces the laser to oscillate in a single longitudinal mode with no intensity beat noise.

4.8 Infrared Countermeasures (IRCM) Systems

IRCM will play a critical role in neutralizing the threats posed by short- and long-range IR missiles and shoulder-fired missiles (SFMs) or man-portable air defense missiles (MANPADMs) known as Stingers. These missiles, when operated by terrorists groups, pose the greatest threat to passenger jets and military aircraft. According to the IEEE Spectrum survey [11], more than 50,000 shoulder-fired missiles are now in existence. This year the U.S. House Appropriations Subcommittee has approved a homeland-security-funding bill authorizing research and development activities on the development of antimissile protection devices for commercial aircraft. One shoulder-fired missile can seriously damage or even bring down a passenger jet carrying 300 or more passengers. The Stingers are easily transportable and can be set up and fired by one person with minimum training. They have an operating range close to 5 km or 16,000 feet. Thus, these missiles pose a serious threat to planes, particularly during takeoff and landing operations.

4.8.1 Threat Assessment and Neutralization of IR Missile Threats

Military aircraft have deployed antimissile defense techniques using decoys such as flares fired from the aircraft or helicopter to confuse the attacking heat-seeking missiles or air-to-air IR missiles. However, an instant and most effective protection is not possible through such decoy devices. The latest IRCM technology offers the most effective and instant protection from hostile heat-seeking missiles. This particular IRCM technology uses compact, diode-pumped, solid-state lasers to ward off deadly IR missiles. The Northrop-Grumman Directional Infrared Countermeasures (DIRCM) system offers guaranteed protection against hostile IR missiles.

4.8.2 Performance Capabilities and Limitations of the DIRCM System

The DIRCM system detects a heat-seeking missile in its launch phase and automatically directs a high-energy laser beam to confuse its guidance electronics. This system hardly weighs 42 kg, needs no maintenance, automatically detects the missile, fends off an attack, and informs the surveillance crew afterward. These systems are very costly and are best suited for deployment in Special Forces fleets consisting of gunships and high-performance fighter/bomber aircraft, whose missions require deep penetration into hostile territory. DIRCM systems are also being developed by aerospace companies in Israel.

The Department of Homeland Security could recommend a modified version of DIRCM for installation in commercial aircraft. The DIRCM system for use in commercial aircraft needs not to meet stringent battlefield performance requirements and could have low-cost maintenance schedules and procedures. According to the latest estimates, such an antimissile defense system will cost between $1 and $2 million. This high cost can be fully justified for a wide-body commercial jet aircraft costing close to $100 million. If all the commercial jet aircraft were equipped with the DIRCM technology, though, they would still be vulnerable to low-tech, cheap weaponry, particularly during the landing and takeoff operations. Relatively "dumb" rocket-propelled grenades (RPGs) (which are essentially large, explosive bullets) can be as lethal as heat-seeking missiles because RPGs have no homing device to be defeated.

4.9 Three-Dimensional Laser Tracking (3-DLT) Systems

A 3-DLT system provides critical measurement tasks in the building, testing, and installation of the Fixed Pad Erector (FPE) system before the launch rocket is transported to its launch pad. The FPE raises the launching rocket vertically, aligns it, and transports it into the launch table. The entire operation is very complex, involving more than two dozen precision interfaces that require tolerances significantly less than 800 microns or 0.08 mm. This laser tracker provides 3-D tracking that is accurate to within 10 +/- 0.8 microns/meter and has an operating range close to 250 feet.

During the construction of the FPE structure, the laser tracker must be set up on the launch pad to measure accurately the interfaces involved [12]. The laser beam is locked onto a handheld retroreflective target, which can be moved by an operator over the surface to be measured. The laser beam is reflected off the target, and its path is retraced to the laser tracker (LT) to measure the distance. Angular encoders are installed at appropriate locations to measure the orientations of the LT's two mechanical axes. Thus, the distance and two angular measurements provide the exact 3-D location of the target. Measured data, as well as 3-D graphics generated from the precision-measured data, can be displayed on a PC monitor or laptop connected to the LT. In summary, the LT pinpoints the exact 3-D location of the FPE and its components at the rocket launch pad.

4.10 Optical Control of Transmit/Receive (T/R) Modules in Phased Array Radars

The recent development of high-speed analog FO links and key circuit topologies offer an effective optical control and interconnects of the microwave monolithic integrated circuit (MMIC) in T/R modules widely used by high-power phased array target tracking and surveillance radars. FO technology plays a key role in MMIC-based systems that require reliable performance under severe operating environments. In addition, the FO technology permits rapid integration of MMIC and photonic components in complex military and space systems [13]. In brief, FO technology provides optical control of MMIC-based T/R modules and is ideal for high-power space tracking radars and phased array missile-defense surveillance radars. Deployment of MMIC-based T/R modules in high-power phased array radars (Figure 4–15) offers multiple target tracking and detection capabilities with high accuracy.

4.10.1 FO Links in Providing Optical Control in T/R Modules

The FO link is a compact and lightweight element. It provides excellent isolation over a wide bandwidth and high immunity to EMI and electromagnetic pulse (EMP). The FO link permits rapid integration of optical-signal processing with fiber interconnects with minimum cost and complexity. However, FO links have drawbacks. They require an interface to the MMIC assembly involving an optical transmitter and receiver. This introduces additional complexity and reduces overall link performance in terms of the gain-bandwidth product, dynamic range, and signal-to-noise ratio (SNR). However, recent developments suggest that

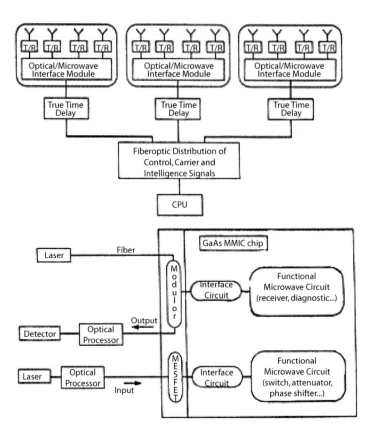

Figure 4–15 *Control signals via FO lines in an optically fed phased array antenna using T/R modules.*

these problems can be overcome if one does not regard the FO link as a direct substitute for the conventional coaxial cable.

4.10.1.1 FO Link Configurations

FO link configurations are classified according to the coding techniques used by the links. Two types of coding techniques can be used: the central coding (CC) technique and the remote coding (RC) technique [13]. In the case of a CC FO link configuration, the link consists of an AlGaAs optical transmitter, a wideband optical receiver, and 50 meters of low-loss optical fiber. In the case of a CC link configuration, an alternative resistive matching is used, and the upconverted signal after mixing with a stabilized IMPATT oscillator, is monitored on a spectrum analyzer to detect the quality of the signal. The CC approach cannot be implemented above K_u-band at present because of high losses in commercially available lasers and detectors at these frequencies. The RC approach has been successfully demonstrated at the 24 and 40 GHz frequency bands by utilizing the nonlinearities of the electro-optic and microwave components. Furthermore, the RC approach uses a reactive matched detector circuit, which offers better performance over the CC FO link configuration, even below 10 GHz operations. In addition, the RC configuration offers a 10 dB better performance in dynamic range than the CC method because of a reactively matched detector circuit [13].

4.10.2 Optical Beamforming

The term "beamforming" applies to electronically steering the RF beam to specified points in azimuth (AZ) and elevation (EL) directions and adjusting the beam-shape parameters, such as sidelobe suppression and 3 dB beamwidth of the phased array antenna through phase shifters and gain controllers. For a phased array radar to operate in complex signal environments, it must be reconfigurable in a rapid, accurate, and efficient manner. This places severe demands on the power and speed of the computational subsystems and requires fast, parallel optical processing techniques.

In an electronically steerable phased array antenna, the excitation to each individual antenna element must be controlled to achieve amplitude tapering to meet sidelobe-level requirements and beam steering in AZ and EL planes. In the case of an optically controlled phased array antenna, the gain and phase parameters for each radiating element are functions of an optical signal. Two broad categories of beamforming techniques can be distinguished, depending on whether the phase shifting and/or attenuation are implemented in the microwave or the optical domain (see Figure 4–16).

4.10.2.1 Beamforming in Microwave Domain

Techniques described under this category use microwave phase shifters, attenuators, and other devices in situ of the MMIC T/R module. The control devices or elements are actuated by control signals produced by GaAs digital circuits and transmitted via optical fibers. The function of the optical fiber link is to transfer the beam control signals from the central processing unit (CPU) to the T/R modules where they are detected, decoded, and utilized to actuate the conventional phase shifting and gain control circuits. In this method, the control wires are replaced by the optical fibers. A monolithically integrated GaAs optical interface chip is deployed to detect 1 Gbit/s multiplexed control/information signals at BER better than 10^{-9} to preserve high phase accuracy. Photodetectors coupled with preamplifiers and buffer amplifiers with high isolation are used to maintain stable performance. These components are interfaced with the digitally controlled 5-bit phase shifter and RF power amplifier in T/R modules (see Figure 4–16).

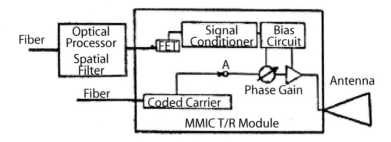

(a) Indirect Optical Control of MMIC-Based T/R Modules

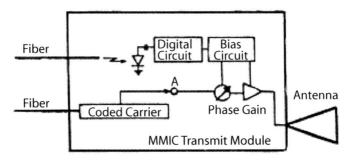

(b) DIrect Optical Control of MMIC-Based T/R Modules

Figure 4-16 *Optically controlled MMIC-based T/R modules using (a) indirect method and (b) direct method.*

4.10.2.2 Beamforming in Optical Domain

This beamforming approach uses an FO network to distribute control signals in conjunction with a parallel optical processing scheme, as shown in Figure 4-16. In a T/R module, the excitation to each antenna element is a function of phase to steer the beam in a specific direction and amplifier gain to provide amplitude tapering for satisfying the sidelobe requirements [13]. In this approach, the amplifier gain and phase of each module are dependent on the optical intensity provided by an LED. The phased array antenna beam pattern can be manipulated and optimized using variable optical spatial filters (see Figure 4-16). An optimum antenna beam pattern requires efficient optical gain and phase controllers. Both the phase-shift control and gain control as a function of optical intensity must provide a linear response over a wide RF bandwidth to meet the wideband capability of the phase array antenna. There are many different ways to implement phase-shifting- and gain-control techniques using optical sources.

Beamforming with FO links is based on true-time delay phase shifts produced on the FO lines connecting the CPU with the T/R modules (Figure 4–15). The true-time delays can be achieved either by changing the effective length of the optical fiber between the CPU and any one of the particular T/R modules, thereby causing a phase shift, or by varying the group velocity of the optical signal carrying the microwave information on the high-speed FO link. Changing the path length on the FO link can be accomplished by switching to different lengths of optical fibers. The distribution of high-frequency and high-data-rate signals requires several narrow microwave beams to track multiple targets. A multiple-target tracking capability in battlefield environments requires beamforming in the optical domain.

4.10.3 Indirect Control of Phase Shifter and Amplifier Gain

In this method, the optical signal is detected, conditioned, and then injected into the gain- and phase-control circuits. Optical control of gain and phase functions in a hybrid GaAs MMIC assembly was successfully demonstrated in 1988 using LED light sources. The gain-control technique consists of four main sections: the LED light source with a pigtail, a light-sensitive field-effect transistor (FET), a DC amplifier, and a distributed microwave amplifier. The hybrid MMIC assembly consists of GaAs MMIC chips. A multifinger metal-semiconductor field effect transistor (MESFET) device is used as a photodetector because of its excellent compatibility with N-type processing of the MMIC. Furthermore, the multifinger MESFET offers more exposed GaAs for light absorption than a single-gate FET and exhibits excellent frequency response up to 1 GHz. Since all the components used are compatible with GaAs MMIC fabrication methods, they could be integrated into a single MMIC chip with minimum cost. The MESFET is biased near pinch-off voltage, where maximum sensitivity to light is obtained. The forward transmission coefficient of the amplifier with flat frequency response can be obtained over 4 to 8 GHz bandwidth for light intensities varying from 50 to 250 microwatts. The gain of the amplifier can be controlled linearly from –10 to +6 dB using 50 microwatts of optical power from the LED source. The amplifier gain control is saturated above 50 microwatts of LED power because the FET acts as an optical detector in the circuit. A 360-degree continuously variable phase shift with 30 dB of gain control is obtained by adjusting the relative gains of two adjacent quadrature vectors in a optically controlled vector-modulator phase shifter. Variable phase values are obtained by modulating the light intensity of the LED source. Dual-gate MESFET devices are used as the variable gain stage using proper gate bias levels. This type of optical control of MMICs can be extended to the mm-wave range and is compatible with emerging optical-signal processing involving spatial filters with potential application in advanced electronically steerable antennas (see Figure 4–16A).

4.10.4 Direct Optical Control Approach

This approach is sophisticated and uses optically controlled microwave devices and subsystems that are incorporated into the T/R module, in which the gain and phase are controlled directly. This is called a direct method because the microwave device—heavily doped p and n regions separated by a lightly doped instrinsic region (PIN) diode or MESFET—acts both as a detector and as a control element. In this method (see Figure 4–16B), the microwave lateral PIN diode is used in both the phase shifter and the attenuator circuits, and the device is amenable to MMIC processing. This approach uses an optically controlled microwave reflection-type phase shifter, which suffers from higher insertion loss. Higher incident optical power levels between 20 to 25 mw will not only eliminate the need for DC bias but also reduce the reflection loss well below 2 dB over DC to 8 GHz microwave region. A lateral PIN diode can be used as an optically controlled microwave switch and a variable attenuator capable of operating from DC to 8 GHz when illuminated by a coherent laser source. An increase in the illumination (from 10 to 25 mw) will result in a uniform drop in the insertion loss across the entire frequency range. Direct optical control of the MMIC-based gain element and oscillator configurations using high performance MESFET, high electron mobility transistor (HEMT) and heterojunction bipolar transistor (HBT) devices must be investigated for further improvement in performance. Studies performed by the author on various solid-state devices indicate that the undoped GaAs region of an HEMT device confined to a two-dimensional electron gas (2-DEG) demonstrates greater sensitivity than the MESFET because of the higher gain provided by the photogenerated carriers.

4.10.5 Control of Passive Elements

Optical control of passive microwave components, such as dielectric resonators and microstrip transmission lines, can be obtained by incorporating a photosensitive material into the microwave element and then actuating the photoconductor by optical source. Light-generated plasma affects the propagation velocity resulting in phase shift and signal attenuation. Photoconductor wafers, when placed on the top of a dielectric resonator, where the photogenerated carriers perturb the electric field configuration in the resonator, act frequency oscillators. Cost-effective optical control of passive microwave components has not yet been demonstrated because of high-intensity optical-source requirements. More research and development activities are needed in this area.

4.11 Summary

This chapter describes the FO-based components and sensors best suited for military and space applications. Performance capabilities and limitations of programmable delay lines widely used in electronic warfare, missile simulation, and radar evaluation are discussed in detail. Performance capabilities of various delay-line configurations using waveguide, coaxial, FO, and acoustic technologies are compared in terms of weight, size, and insertion loss. Performance requirements for critical elements of the FO-based delay line are defined. Advantages of an FO Michelson array that is best suited for passive elimination of polarization fading are summarized. Performance capabilities and selected design configurations for various tunable dispersion compensators are described. Critical performance parameters of a tunable dispersion compensator, including quadratic group delay, tuning range, dispersion slope, and spectral bandwidth are discussed in detail. FO-based ring gyros, which have potential applications in military weapon systems and space sensors, are described with emphasis on modulating light-source intensity and Kerr-effect-induced bias error. All-fiber Q-switched lasers widely used in LIDARs, rangefinders, and distributed sensors are briefly discussed. Performance requirements of a solid-state laser illuminator, which is a critical element of the beam and fire control systems in a high-power airborne laser, are summarized. The impact of atmospheric parameters on turbulence intensity and optical wavefront error is reviewed with emphasis on volumetric density of aerosols in various atmospheric layers. An effective IR countermeasures system capable of neutralizing the threat posed by hostile, shoulder-fired heat-sinking missiles is described. Operating range and destructive power of a shoulder-fired missile are briefly discussed. Three-dimensional laser tracking equipment is described with emphasis on interface tolerance requirements and its operating range for a rocket launch pad. FO-based optical-control techniques for MMIC-based T/R modules used in a phased array antenna are described with emphasis on phase-shifter- and amplifier-performance parameters as a function of light-source intensity. Critical elements of a T/R module are briefly discussed.

4.12 References

1. Jha, A. R. (1978, September). *Programmable fiber optic delay lines* (technical report, pp. 5–18). Jha Technical Consulting Services.

2. Marrone, M. J., et al. (1991). Polarization-intensity fiber optic Michelson interferometer. *Electronic Letters, 26,* 518.

3. Fells, J. A. J., et al. (2001). Twin-fiber grating tunable dispersion compensator. *IEEE Photonics Technology Letters, 13*(9), 984–986.

4. Williams, J. A. R., et al. (1991). Fiber Bragg grating fabrication for dispersion slope compensation. *IEEE Photonics Technology Letters, 8,* 1187–1189.

5. Hotate, K., et al. (1991, September). *Fiber optic gyros.* Invited to 15th Annual Conference, Boston, MA, pp. 1585–1614.

6. Hitz, B. (2003, October). Tunable dispersion compensator uses micromirrors. *Photonics Spectra*, 23–24.
7. Jha, A. R. (2000). *Infrared technology: Application to electro-optics, photonic devices, and sensors* (pp. 120–124). New York: John Wiley and Sons, Inc.
8. Johnson, B. D. (2003, May). Airborne laser illuminator delivered. *Photonics Spectra*, 39.
9. Jha, A. R. (2000). *Infrared technology: Application to electro-optics, photonic devices, and sensors* (pp. 58–60). New York: John Wiley and Sons, Inc.
10. Senior Editor. (2003, July). Eliminating beat noise in ring gyros. *Photonics Spectra*, 104–105.
11. Kariya, S. (2003, August). Missile defense for airlines finds growing gupport in United States. *IEEE Spectrum*, 14–15.
12. Budimir, M. (2003, September). Laser measurements aids rocket launch. *Machine Design*, 35.
13. Herczield, P. R., et al. (1998, May). Optical control of MMIC-based T/R modules. *Microwave Journal*, 309–320.
14. Kunath, R. R., et al. (1986). Optically controlled phased array antennas using GaAs MMICs. *IEEE AP-S, International Symposium Digest, 1*, 353–357.

CHAPTER 5

Integration of Fiber Optic Technology in Commercial and Industrial Systems

Recent advancements in FO technology promise effective integration of FO devices and components in commercial and industrial systems. FO components can play a key role in the design, development, and manufacturing of industrial and commercial machines, including automobiles, tractors, trucks, production-monitoring sensors, and communication equipment. The explosive growth of internet protocol (IP) traffic and voice-communication channels requires increased access to high-speed Internet connections by businesses and local and overseas customers. FO technology offers not only high-speed data and voice transmission, but also provides very efficient and reliable service with minimum cost. Critical component requirements for possible use in code-division multiple-access (CDMA) base stations, time-division multiple-access (TDMA) subscriber stations, automobile manufacturing, gyroscopes for aerospace sensors, production quality-control systems, and industrial sensors for detection of poisonous gases and chemical agents will be identified. Requirements for critical components used in various commercial and industrial optical sensors will be summarized with an emphasis on performance, cost, and reliability while in harsh operating environments. Optical components such as switched, multiport-circulators and fiber Bragg gratings (FBGs) will be briefly discussed with an emphasis on performance capabilities and limitations.

5.1 Optical Sensor Using Cryogenic and FO Technologies

Optical spectroscopic sensors using both the cryogenic and FO technologies described here play a critical role in the research and development of rare-earth compounds. High-resolution spectroscopy is a powerful tool for investigating the interactions in ionic rare-earth compounds most suited for aerospace applications. This equipment provides valuable information on the energy of the electronic crystal-field states of the ions; their eigenfunctions; and their magnetic moments as a function of temperature, magnetic field, and pressure. Investigation of cooperative interactions and magnetic-field and structural-phase transitions finds that they normally do not require liquid-helium cooling. But the noncooperative crystal-field interactions need liquid-helium cooling to meet high-resolution requirements. The resolution of the spectroscopic sensor or equipment is dependent on cryogenic cooling, optimum optical fiber parameters, and the widths of the entrance and exit slits. Performance trade-off studies indicate that these widths should not exceed 50 microns if high resolution is the principal requirement. This sensor is best suited for automatic data recovery and processing in complex space applications [1].

5.1.1 Description of the Critical Elements and Their Performance Requirements

Critical elements of this optical equipment include a sample holder, a heater, a cryogenic chamber with lid, and input and output FO lines, as shown in Figure 5–1. The sensor construction must be such that the heating of the helium chamber due to irradiation from the chamber surroundings and the incident light must be minimized. Optical spectroscopes are usually equipped with FOBs to retain their transparency at cryogenic temperatures. Conventional optical fibers using a quartz core and quartz cladding tend to loose their transparency at a cryogenic temperature of 100 K. However, an optical fiber consisting of a quartz core and polymer cladding maintains excellent transparency down to a cryogenic temperature of 0.68 K [1].

The attenuation for these particular optical fibers should be less than 45 dB/km or 0.045 dB/meter at room temperature (300 K) in the visible spectral region. Laboratory tests exhibit acceptable insertion loss even at 4.2 K and coupling loss of less than 0.2 dB for each of the four coupling mechanisms. Single optical fibers with a cladding diameter of 140 microns and a core diameter of 125 microns offer greater effective areas. The core diameter is restricted to 125 microns to maintain a bend radius not to exceed 20 mm. Four such optical fibers are used to guide the light energy from the slit of the spectroscope. After the exit slit, the four fibers are mounted one upon another in a line. Since the polymer coating is weaker than the quartz, the core diameter of each fiber may look convex.

5.1.2 Performance Requirements for Helium Cryostat and Nitrogen-Storage Tank

The storage tank for liquid nitrogen (N_2) is covered with ten layers of superinsulation to maintain high cryogenic-cooling efficiency. The storage tank has a radiation shield in the upper part, which is in thermal contact with the nitrogen bath via eight copper rods. The storage tank can stay cool for four days. The liquid helium (4He) has capacity to retain the liquid for a maximum of two days, which is strictly limited by the thermal conductivity of the leads of the magnet. The neck of the helium tank contains Styro-based insulation. Between the two tanks, there is a helium-cooled radiation shield so that helium vapor can escape. The superconducting magnet is located in the lower part of the He tank. The test

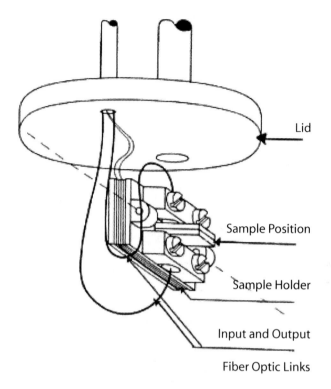

Figure 5–1 *Elements of the experimental chamber including the chamber lid of the cryogenically cooled optical spectroscope.*

chamber is flooded with either liquid helium (4He) or gaseous helium (3He) for cooling the condensation. The lowest temperature that can be reached is 1.05 K with liquid helium or 0.68 K using gaseous helium.

5.1.3 Sample Requirements and Test Procedures

The sample holder is suspended from the lid of the chamber as illustrated in Figure 5–1. The sample adjustment and orientation are required in an optimum position relative to the magnetic field to achieve reliable test data. The optical fibers carrying the input and output signal are fixed to two copper strips as shown in the figure. The sample to be investigated lies between thin sheets with holes serving as a diaphragm. To insert the sample into the sample holder, the experimental chamber (not shown in Figure 5–1) must be taken out of the cryostat. The lower portion of the experimental chamber, which is fixed to the lid, must be removed. The sample can then be attached to the holder, and its orientation can be corrected simply by turning the holder assembly. The holder is fastened in the desired position and then mounted in the cryostat for testing.

The desired operating temperature is selected and maintained for a specific duration after the liquid helium is pumped in the experimental chamber (not shown) at a rate required to reach the selected temperature. The temperature must be computer controlled to keep the chamber temperature at the required level to within 1 mK of the cryogenic temperature range up to 4.2 K.

5.1.4 Performance Capabilities of the Optical Spectroscope

Based on test procedures, required measurements can be made for approximately 30 minutes at a temperature of 0.7 K with one condensation of the helium gas. Furthermore, it is easy to achieve the light intensity or signal-to-noise ratio (SNR) at the photomultipler using single optical fibers. This cryogenically cooled optical spectroscope permits inhomogeneity measurements of a sample with a very small cross-section by focusing the light on its different parts.

The configuration of an optical spectroscope described here is best suited for spectra measurements of the rare-earth compounds such as erbium-oxi-sulphate ($Er_2O_2S0_2$) as a function of temperature and magnetic field. The spectra measurements offer precise determination of the magnetic structures in various magnetic phases.

5.2 All-Optical Nonlinear (AON) Switch

An ultrafast all-optical switch is critical in fiber communications and telecommunications systems. The most promising nonlinear switching scheme is based on nonlinear optical loop mirrors (NOLMs). Note loop mirrors rely on nonlinear face shift, which switches the state interference from destructive to constructive and vice versa. The interferometric characteristic of an NOLM makes it extremely sensitive to vibration, polarization, and optical pump output power fluctuations. In cross-phase modulation (CPM) the evolution of the signal spectrum can be controlled by selecting the right pulse width, proper initial delay, and optimum walk off between the signal and pump pulses. A nonlinear switching scheme using the property of CPM and an FBG device offers efficient switching action, minimum loss during "ON" position, and robust structure. This frequency-shifting switching scheme is not affected by interference and does not require precise values of nonlinear phase shift.

5.2.1 Design Aspects of the Switch

A schematic diagram of the AON switch is shown in Figure 5–2. The AON switch consists of an optical fiber of length L with an FBG in the middle of the fiber length. In the OFF position, a weak signal pulse propagates without any change from location $z = 0$ to $z = L/2$, where it is completely reflected by the FBG device acting as a bandstop filter.

In the ON position, the optical pulse propagates along with the strong pump pulse also known as the control pulse. The amplitude of the pump pulse is five to 10 times larger than that of a signal pulse. The control pulse shifts the signal spectrum by CPM and allows it to bypass the FBG device [2].

5.2.1.1 Mathematical Expressions for Frequency Shift and Nonlinear Phase Shift

The instantaneous nonlinear (NL) cross-phase modulation frequency shift can be written as

$$\Delta w_{NL}(\tau, z) = [2\beta_s n_2 / w][I_p(\tau - wz) - I_p(\tau)] \quad 5.1$$

where Ip is pump-pulse intensity, w is the walkoff parameter, τ is the signal pulse width, n_2 is the NL refractive index of the fiber core, and β_s is the propagation constant for the signal wave.

The signal pulse can be defined as

$$\tau = [t - v_s z] \quad 5.2$$

where v_s is the group velocity at the signal wavelength and z is the propagation distance. The refractive index (n) of the fiber has two components: the linear component (n_0) and NL component (n_2). The parameter n is a function of τ and ε which is equal to distance z. Thus, the overall refractive index of the optical fiber can be written as

$$n(\tau, \varepsilon) = [n_0 + 2n_2 I_p(\tau, \varepsilon)] \quad 5.3$$

The NL phase increment can be given as

$$[d\phi_{NL}(\tau, \varepsilon)] = [2\beta_s n_2 I_p(\tau, \varepsilon) d\varepsilon] \quad 5.4$$

Assuming the pump-pulse shape remains unchanged during the propagation, the pump-pulse intensity can be written as

$$I_p(\tau, \varepsilon) = [I_p(\tau - w\varepsilon)] \quad 5.5$$

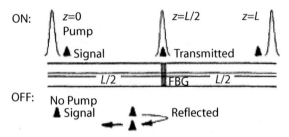

(a) Nonlinear Pulse Switching Scheme Showing the Signal and Pump Pulses

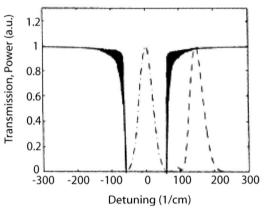

(b) Spectra of Signal Pulse in OFF (Dotted) and ON Positions (Dashed) and Transmission Spectra of Step-Chirped Grating (Solid Line)

Figure 5–2 *(a) Signal and pump pulses and (b) transmission spectrum.*

The NL phase shift produced by the point of the signal pulse over a propagation distance equal to z is proportional to the integral of the pump-pulse shape and can be expressed as

$$\Delta\phi_{NL}(\tau,\varepsilon) = [(2\beta n_2)\int I_p(\tau - w\varepsilon)d\varepsilon] \qquad 5.6$$

where the integration is carried out from 0 distance to z distance. The instantaneous frequency shift is the time derivative of the above equation, which is represented by the Equation 5.1.

5.2.2 Conditions for Distortion-Free Signal Transmission

To achieve the most efficient switching action, i.e., complete and distortion-free signal transmission through the optical switch in the ON position, one must make the instantaneous nonlinear frequency shift as large as possible. This means the quantity $[\Delta w_{NL}(\tau, L/2)]$ must be maximized while keeping the quantity $[\Delta w_{NL}(\tau, L)]$ close to zero for all points τ within the signal duration. The way to satisfy these conditions is that the initial delay (T_d) between the peaks of the signal and control pulses not permit overlap at $z = 0$ and $z = L$, but coincide at $z = L/2$, as shown in Figure 5–2.

These two specific conditions can be satisfied by selecting the time delay between the signal and control or pump pulses, which is expressed as

$$T_d = [wL/2] \geq [\tau_s + \tau_p]$$

5.7

where symbol τ indicates the pulse duration and subscripts s and p represent signal and pump pulses. Using Equations 5.6 and 5.7, Equation 5.1 can be rewritten as

$$[\Delta w_{NL}(\tau, L/2)] = [(2\beta n_2 / w) I_p (\tau - \tau_p)]$$

5.8

and also

$$[\Delta w_{NL}(\tau, L)] = 0$$

5.9

It is evident from these equations that the frequency shift must be large enough to locate the signal completely outside the filter stopband region. To reject the signal in OFF position, the stopband of the FBG-based filter must be greater than the bandwidth of the signal, which can be mathematically expressed as

$$[\Delta w_{stop}] \geq [\Delta w_s] \approx [2\pi / \tau_s]$$

5.10

5.2.3 Pump Power Requirements

The pump requirement is directly proportional to the pump power-length product and inversely proportional to the propagation constant, refractive index of the fiber core, and pump length. The pump power-length product can be expressed as

$$P_{PL} = [(1/A_e)(2\beta n_2 L P_{pump})] \qquad 5.11$$

where A_e is the effective area of the optical fiber and P_{pump} is pump power, which is the product of pump pulse intensity (I_p) and the effective area of the fiber (A_e).

When the length z is equal to $L/2$ or in the middle of the fiber, it is assumed the pump intensity is maximum all over the signal-pulse duration.

This means the approximate expression for the pump intensity can be written as

$$I_p = [I_p(\tau_s - \tau_p)] \qquad 5.12$$

where τ_p is the pump-pulse duration or control-pulse duration. The maximum pump intensity occurs when the pump-pulse duration is significantly longer than the signal-pulse duration, i.e., when p is greater than or equal to $3\tau_s$.

The FBG filter located in the middle of the switch must reflect the signal in the OFF position and allow the signal to pass through the ON position of the switch. Deployment of the chirped gratings offers significant reduction in the reflected sidelobe levels and out-of-band dispersion.

5.2.4 Benefits of Chirped FBGs

Use of chirped FBGs offers lower reflection sidelobes and out-of-band dispersion. Based on switching of 3 ps pulses with full width at half-maximum (FWHM) amplitude, the signal spectra at the filter in OFF and ON switch positions along with the transmission spectrum of a chirped grating are clearly illustrated in Figure 5–2. Spectra of the signal pulse in the center of the switch in OFF and ON positions and transmission spectra of the step-chirped grating are based on the grating parameters.

The grating consists of five uniform sections of coupling constant $k = 12$ cm and grating length $L_{grat} = 6$ mm each, with suppression of the pulse spectrum by 30 dB. The grating suppresses the input signal by 30 dB in the OFF position, while experiencing a signal loss of only 0.04 dB in the OFF position. It is possible to design a FBG filter with lower reflection sidelobe levels and out-of-band dispersion by using a grating apodization technique, different chirp profiles, and optimum grating parameters. Chirp grating with a small coupling constant, a large grating length, and a fixed total stopband region will yield lower sidelobe levels.

5.2.5 Performance Capabilities of the NL Pulse Switch

The switch described here employs optical fibers of 1 km length, signal-pulse duration of 3 ps, pump-pulse duration of 10 ps, signal and pump wavelength of 1576 nm and 1530 nm, respectively, and a peak pump power of 4.5 W. The signal spectrum is shifted as the signal and pump pulses walk through each other with a maximum shift in the middle of the switch. This permits the signal to bypass the filter and return to the original spectrum with no pulse dispersion. In the ON switch position, more than 99% of the pulse energy is transmitted. In the OFF switch position, there is no frequency shift and the signal is reflected by the FBG filter, allowing transmission of only 0.1% of the energy. The reflected pulse is broadened by the use of chirped FBG. In the case of two chirp FBG filters, the bit interval of 14 ps is sufficient to allow complete switching of the signal pulses without affecting the adjacent pulses. The switch in this configuration can operate successfully at a bit rate of more than 70 Gbits/s (bit rate = 1/14 ps = 70 Gbit/s).

The effect of dispersion on the switch operation is important for the pump pulse because its spectrum is broadened through self-phase modulation due to NL phase shift. Spectral broadening along with the fiber dispersion causes distortion in the pump pulse, which does not affect the switch efficiency as long as the pump broadened line width ($\Delta\lambda_p$) is much smaller than the difference between the pump and signal wavelength. Furthermore, if the pump pulse is not distorted too much, the switch can operate without any penalty. Reduction of dispersion requires a dispersion-shifted optical fiber with small dispersion slope and the selection of the pump wavelength to coincide with the zero-dispersion wavelength ($\lambda_p = \lambda_0$).

The switch can operate over a wide range of signal wavelengths if the initial delay between the signal and pump or control pulse can be given as

$$[2T_d / L] = [D(\lambda_s - \lambda_o)^2 / 2] \qquad 5.13$$

where D is the dispersion (ps/nm-km), L is the fiber length, and λ_o is zero-dispersion wavelength. Higher nonlinearity-to-dispersion ratios permits lower switch lengths and lower switching power levels. Pulse duration is limited by group-velocity dispersion for the signal and the zero order-dispersion for the pump waves. One way to reduce pulse duration is to reduce the switch length (L) and increase the pump power. NLOMs, which are widely used in advanced NL optical switches, require cross-phase modulated phase shifters.

The cross-phase modulated phase acquired by the signal pulse can be expressed as

$$\Phi_{NOLM} = [2\beta n_2 LP_p / A_e] = [\pi] \qquad 5.14$$

In the case of an NOLM switch, large interferometer arm lengths are required, but the large arms suffer from large distortion and dispersion. In the case of NL pulse switching (NPS) schemes using CPM and FBG techniques, the pump power-length product is smaller than that for a conventional NOLM by a factor of four times the sum of signal and pump pulse durations divided by the signal pulse duration.

With short signal pulses, an NL FBG requires a pump power intensity (I_p) close to 600 kW, which is five orders of magnitude compared to a switching scheme compromised of CPM and FBG. In summary, the AON switching (AONS) scheme is more robust and provides the most efficient and fast switching action over a well-designed NOLM-based switch, strong signal rejection in the OFF state, and practically zero transmission loss in the ON state. The lowest distortion both in time and frequency domains can be achieved by using strained fiber grating filters and narrow pump pulses. However, the AONS performance is slightly limited by the fiber dispersion at elevated temperatures and high pump power requirements in the case of switching narrow signal pulses. This AONS, with very high ON/OFF contrast and practically no pulse distortion, is best suited for long-haul communications and telecommunications systems requiring data rates as high as 75 Gbits/s.

5.3 Optical System for Tracking High-Speed Train Tracks

According to the latest survey on train accidents in the USA, a train accident occurs every 90 minutes [3]. Most of these accidents are attributable to undetected cracks in the tracks. Ultrasonic techniques are currently used to investigate the mechanical integrity of the tracks, but they are limited in application and have real trouble in detecting the cracks that are vertical to the top surface of the rail or in its base structure. A laser-based ultrasound system (LBUS) described here detects flaws missed by the instruments currently used and completes the entire operation with remarkable speed and reliability.

5.3.1 Description and Operating Principle of LBUS

The LBUS uses a Q-switched Nd:YAG laser source to generate ultrasonic waves in the rail by localized ablation of the rail surface. The laser provides 400 mJ of optical energy at 1064 nm radiation with a pulse width ranging from 4 to 8 ns, which is delivered to the rail through an optical fiber. This LBUS system incorporates a capacitive air-coupled transducer to monitor the frequencies from 50 to 2 MHz. The transducer is the most critical element of the system to monitor cracks of the greatest concern to the railroad industry. The system performance is strictly dependent on the reflectivity of the rail surface that is under investigation. With a single laser pulse, longitudinal, shear, and surface modes propagate in the rail material. This technique detects cracks at almost any inclination, including the surface, and internal cracks on the base of the rail. One can deploy magnetic induction and conventional ultrasound systems, but they require physical contacts with the rail surface, which pose serious logic problems. The LBUS system requires no contact with the rail surface because the capacitance air-coupled transducer can afford some distance between the detection sensor and the rail surface.

The wheel of the system is turning and rolling across the track with no physical contact with the rail. These FO-based systems need to be placed so that when the wheel is in the desired position, mirrors and lenses incorporated in the system come to place and a laser is triggered to generate nanosecond pulses. The air-coupled transducer needs to be placed in such a position that an optical signal reflected from the rail surface is detected. Nothing can obstruct the path of the wheel, and the flip-flop motion is to be repeated for every passing wheel. The reflected signal amplitude can be calibrated in terms of crack width in the rail surface or in the base of the track. In summary, this FO-based sensor detects cracks in the rail surfaces and warns the train operator about the danger ahead. This sensor can play a key role in avoiding train accidents.

5.4 FO Gyroscope (FOG) for Aerospace Applications

An FOG is a solid-state, compact, lightweight device that has potential application in military and space-based systems. As a matter of fact, this device is best suited for multiple aerospace applications. The sensing loop of this device is made from a PM fiber, which is more expensive than an ordinary SM fiber. A depolarizer is inserted at one end of the coil, as illustrated in Figure 5–3. The FOG is very reliable even under severe operating environments.

5.4.1 Design Aspects

The sensor performance is contingent on the interference between the X and Y components of polarization in the coil. Precision performance from this device is not possible due to its very small phase difference between the X and Y polarization components. However, a gyro with low drift can be designed using a coherent light source, such as a super-luminescent diode (SLD) [4]. The required phase difference between the two polarization components can be written as

$$\phi_{pd}(L_c, \lambda) = [(L_c / \lambda) 2\pi] \qquad 5.15$$

where L_c is the coherent source length and λ is the source wavelength. The coherent length of the optical source is defined as

$$L_c = [\lambda^2 / \Delta \lambda] \qquad 5.16$$

where $\Delta\lambda$ is the spectral width of the light source. A Lyot-type depolarizer is generally used to produce the required phase difference (ϕ_{pd}). Some gyro errors can be expected due to deployment of different kinds of optical fiber at one end of the coil, as shown in Figure 5–3. The use of a depolarizer provides a thermally induced nonreciprocity similar to an FO interference version. By providing a depolarizer at an appropriate location (see Figure 5–3a), the phase error ($d\phi_{pe}$) produced in an interval of dl can be written as:

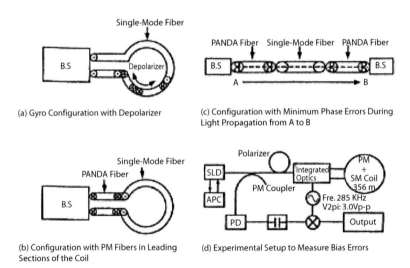

Figure 5–3 *Various gyro configurations: (a) with depolarizer, (b) for minimum phase errors, (c) with PM fiber in leading sections of coil, and (d) experimental setup to measure bias errors.*

$$d\phi_{pe} = [(d\beta/dT) + \alpha\beta][(\delta T/\delta t)(\tau dl)] \qquad (5.16\text{ A})$$

where β is the propagation constant, α is the linear coefficient of thermal expansion in the longitudinal direction or fiber axial direction, τ is the elapsed time after the clockwise light beam has traveled over the distance interval dl until the counterclockwise light transverses the same interval, T is the temperature of the fiber, t is the time duration, and δ represents the change in the parameter involved. The elapsed time is defined as

$$\tau = [n(2l - L)/c] \qquad 5.17$$

where l is the length of the PM fiber, L is the length of the SM-fiber coil and c is the velocity of light. Assuming n_{PM} as the refractive index of the PM fiber core, n_{SM} as the refractive index of the SM-fiber core, ϕ_1 as the phase error produced at the depolarizer, and ϕ_2 as the phase error produced at a symmetrical position on the center of the fiber length, the expression for the difference between these phase errors ($\Delta\phi$) can be obtained. Using Equations 5.16 and 5.17, one can write

$$d\phi_1 = [k(dn_{PM}/dT + \alpha n_{PM})][(\delta T/\delta t)\{n_{SM}(2l - L)/c\}] \qquad 5.18$$

$$d\phi_2 = [k(dn_{SM}/dT + \alpha n_{SM})][(\delta T/\delta t)\{n_{SM}(2l - L)/c\}] \qquad 5.19$$

$$\Delta\phi = [d\phi_1 - d\phi_2] \qquad 5.20$$

$$= (kn_{SM}/c)[dn_{PM}/dT - dn_{SM}/dT + \alpha(n_{PM} - n_{SM})][(\delta T/\delta t)\{(2l - L)/c(dl)\}] \qquad 5.21$$

The first term in the large parenthesis indicates the change in the refractive index as a function of temperature, the second term ($\delta T/\delta t$) indicates the temperature change rate, and the third term $\{(2l-L)dl\}$ indicates the point where the phase errors occur.

By locating the depolarizer at the center of the coil, i.e., when l is equal to $L/2$, the phase errors can be reduced to zero. However, such a condition will degrade the design flexibility of the system. To avoid this problem, the system should not deploy a Lyot-type filter, as shown in Figure 5–3b. On the contrary, a polarization module (PM) fiber is wound into each leading section of the coil, as illustrated in

Figure 5–3b using a PANDA fiber or PM fiber. In this way, the light propagates from a fiber of low birefringence (i.e., SM fiber) to the fiber of high birefringence (i.e., PM fiber). A reasonable polarization can be maintained, but one can expect polarization fluctuations due to environmental perturbations. Connecting PM fibers of equal length to both ends of the coil results in better reciprocity and significantly reduced thermal reciprocity [5].

The most critical system-design parameter is the length of the PM fiber to be connected to the coil. The minimum phase difference between the X and Y polarization components generated during the light propagation from point A to point B (see Figure 5–3c) is given as

$$\Delta\phi = [(\beta_{SMX} - \beta_{SMY})l_{SM} - 2(\beta_{PMX} - \beta_{PMY})l_{PM}] \qquad 5.22$$

where l_{SM} indicates the length of the SM fiber, l_{PM} indicates the length of the PM fiber, β is the propagation constant for the fiber involved, and X and Y are the polarization components along X and Y direction.

The phase difference under the stated conditions is sufficiently large so as not to cause any interference, considering the coherent length (L_c) of the light source involved. The coherent length of the source is given as

$$L_c = [\lambda^2 / \Delta\lambda] \text{ meters} \qquad 5.23$$

where λ is the source wavelength and $\Delta\lambda$ is the spectral width of the source, which is typically between 0.5 and 2 nm.

The phase difference due to the coherent length is expressed as

$$\phi_d = [(L_c / \lambda)(2\pi)] \text{ radian} \qquad 5.24$$

Inserting Equation 5.23 into Equation 5.24, the phase difference can be written as

$$\phi_d = [(\lambda / \Delta\lambda)2\pi] \text{ radian} \qquad 5.25$$

These computations indicate that the phase difference values are sufficiently large (see Table 5–1). The sufficiently large phase requirement can be satisfied if

$$\Delta\phi \geq [(L_c / \lambda)(2\pi)] \qquad 5.26$$

Table 5–1 *Computed values of coherent length and phase difference.*

Spectral Width (nm)	Coherent Length (meter)		Phase Difference (radian)	
	1550 nm	1310 nm	1550 nm	1310 nm
0.5	0.0048	0.0034	19,478	16,462
1.0	0.0024	0.0017	9,739	8,231
1.5	0.0016	0.0011	6,492	5,487
2.0	0.0012	0.0008	4,869	4,116

Using Equation 5.26, Equation 5.22 can be rewritten as

$$[(\beta_{SMX} - \beta_{SMY})l_{SM} - 2(\beta_{PMX} - \beta_{PMY})l_{PM}] \geq [(L_c/\lambda)(2\pi)] \qquad 5.27$$

Dividing each term of this equation by $2(\beta_{PMX} - \beta_{PMI})$, one gets

$$[l_{PM}] \leq [(\beta_{SMX} - \beta_{SMY})l_{SM}/2(\beta_{PMX} - \beta_{PMY})] - [(2\pi L_c/\lambda)/2(\beta_{PMX} - \beta_{PMY})] \qquad 5.28$$

An FOG with a specified precision or accuracy can be designed if a PM optical fiber filter of length l_{PM} satisfying the Equation 5.28 condition is connected to each end of the coil made from an SM fiber, as illustrated in Figure 5–3c.

5.4.2 Critical Performance Parameters

The accuracy of the FOG is strictly dependent on the bias stability for the depolarizer fiber, PANDA fiber, and SM-fiber coil. The bias stability is a function of time constants for each of the three elements involved. The configuration shown in Figure 5–3d can determine the bias stability of any of the three elements; depolarized fiber, PANDA fiber, and SM-fiber coil. The typical time constant for each of them varies from 8 to 10 seconds. The FOG output signals are delivered through an open loop using a lock-in amplifier (LIA), as illustrated in Figure 5–3d. The bias stability of the FOG configuration shown in Figure 5–3d, with a depolarizer at one end of the coil, is roughly 18° C/hr over the temperature range of –20° C to +70° C. The bias stability of an FOG with a PM fiber or PANDA fiber in the coil is less than 9° C/hr over the same temperature range. However, the bias stability of the coil made from an SM fiber is less than 6° C/hr over the same temperature range. The bias stability of the PM-fiber coil and SM-fiber coil is roughly the same at a temperature rate of change of 0.5° C/min. For the two PM-fiber and SM-fiber coils the bias stability is insensitive to time-dependent perturbations as a function of temperature change. In addition, bias periodical fluctuations can occur in the case of a test configuration using a depolarizer. Performance trade-off studies indicate that the FOG configuration using an SM fiber offers a good overall performance with minimum cost and complexity. However, its cost can slightly increase if high precision or accuracy is the principal design requirement. For most aerospace applications with marginal precision, the low-cost FOG configuration described here is quite adequate.

5.5 FO Voltage Sensor (FOVS)

An FOVS provides a safe FO technique for measurement of high-voltage (HV) intensity in electric transmission and distribution lines. The optical instrumentation technique described here has several advantages over the current conventional techniques in terms of bandwidth, isolation, dynamic range, and immunity to electromagnetic (EM) interferences. The FO-based sensor is very small in size, highly reliable, and can be manufactured at minimum cost. More significantly, any sensor failure does not present a danger to electrical-power-station personnel or equipment, in contrast to conventional oil-filed transformers that can explode and completely interrupt electrical service to customers. Optical sensors using the electro-optic effect in the bulk crystalline materials and an elliptical-core, dual-mode fiber for transducer interrogation do not provide high sensitivity and reliable operation.

The FO-based interferometric sensor shown in Figure 5–4 offers high sensitivity and remarkable resolution [6]. However, the differential drift in the arm produced by the random fluctuations as a function of environmental parameters such as pressure and temperature degrade the resolution of the sensor at low frequencies. In addition, this sensor can suffer from inaccurate polarization control and polarization-induced fading. The problems due to phase shift and polarization-induced fading do not seriously degrade the sensor performance because the sensor measures the optical-signal parameters rather than the signal amplitude. In addition, the sensor described here deploys a unique frequency-coding scheme that permits measurement of the spectral characteristics of the optical signal rather than signal amplitude. The decoding scheme used by the sensor allows the estimation of the AC voltage amplitude by measuring the spectral bandwidth of the interferometer output signal.

5.5.1 Critical Elements

The Mach-Zehnder interferometer (MZI) is the most critical element of the optical sensor shown in Figure 5–4. The FOVS is powered by a distributed feedback (DFB) laser with an extremely narrow spectral bandwidth close to 0.1 nm, which can yield a coherent length of 0.024 meters at a wavelength of 1550 nm and 0.017 meters at a wavelength of 1310 nm. Coherent-length computations as a function of operating wavelength and spectral width reveal that a coherent length of 0.024 is required to achieve one-meter accuracy at 1550 nm and a coherent length of 0.017 is required to achieve the same accuracy at 1310 nm. The computations further reveal that a narrow spectral bandwidth requires a shorter interferometer-arm length to achieve the same one-meter accuracy.

The output of the interferometer can be measured with a 125 MHz detector, shown in Figure 5–4. The interferometer deploys two polarization controllers (PC1 and PC2) that require no adjustments and are used only to avoid null visibility of the fringes. A telecommunications-grade SM fiber (Corning SMF-28) with a length of 38.3 meters is wound onto the main piezoelectric tube (PZT1) to meet the one-meter accuracy, assuming the laser wavelength of 1615 nm and spectral width of 0.1 nm. The SMF-28 optical fiber is best suited for CATV and telephony applications. It is also widely used for long-haul communications lines, feeder lines, and distribution networks carrying voice, data, and video services. A second piezoelectric tube (PZT2) with a length of 36.9 meters of the same fiber is wound onto the second arm of the interferometer. The piezoelectric tubes (PZTs) are made from high performance ceramics. The length, diameter, and wall thickness of each PZT are 7.62 cm, 5.08 cm, and 0.41 cm, respectively. Two equal arm lengths of the interferometer are required to achieve the accuracy of +/- 1 meter.

5.5.2 Performance Parameters

The output of the interferometer is a frequency-modulated signal. When a harmonic electrical signal with specific amplitude is applied to the PZT, the optical fiber wound is strained and a change in phase occurs, which can be written as

$$\Delta\phi = [\phi_0 + \phi_{peak} sin(wt)] \qquad 5.29$$

where ϕ_0 is the phase at $t = 0$ and ϕ_p is the phase at the peak applied voltage Vp. Note the phase modulation is proportional to the applied voltage.

The frequency spectrum of the interferometer output can be expressed in Fourier series involving Bessel functions, whose argument is ϕ_p. The bandwidth of the interferometer is controlled by the phase at applied voltage. The voltage amplitude Vp can be obtained by measuring the spectral width ($\Delta\phi$). The amplitude of the applied voltage can also be obtained by counting the number of fringes of the interferometer output signals. The maximum frequency is proportional to the applied voltage. The PZTs exhibit linear behavior, and the efficiency (η_s) of the sensor in terms of fringes

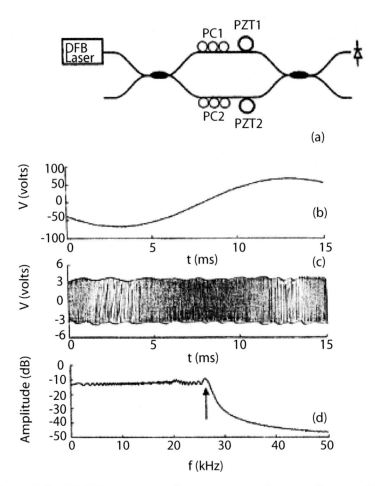

Figure 5-4 *FO HV transmission line sensor: (a) schematic diagram of MZI, (b) output voltage response, (c) modulated output of interferometer, and (d) frequency spectrum.*

is better than 3.14 fringes per volt of the applied voltage, which yields the modulation efficiency (η_m) close to 6.98 radians/volt. In both PZTs, the relative deviations from the linearity with respect to best linearity fit are within +/-1% for applied voltages exceeding 25 volts. The calibration of PZT2 yields a modulation efficiency of 6.50 radians/volt. The voltage sensitivity of this sensor can be given as

$$\Delta V = [2\pi / \eta_m] \text{radian / volt} \qquad 5.30$$

The voltage sensitivity can be improved by increasing the modulation efficiency, which can be achieved by coiling a longer length of fiber onto the PZT. For example, a coil consisting of 375 meters of SMF-28 fiber yields a modulation efficiency close to 74.2 radians/volt.

When two AC signals are applied simultaneously to the two PZTs, the phase difference ($\Delta\phi$) is now given by the difference between the two phase shifts generated ϕ_1 and ϕ_2. These are proportional to V_1 and V_2 voltage, respectively. The interferometer output is an optical signal that can be correlated to the subtraction of the two voltages. If the two phase modulations are balanced, i.e., both exhibit the same modulation efficiency, the output of the sensor will be a straight line. When the same voltage is applied to both PZTs, the modulation efficiency is 0.48 radians/volt, which agrees with the difference between the two transducers' efficiencies. The FO-based interferometer sensor for HV sensing or monitoring is insensitive to phase drift and polarization fading. This sensor is best suited for monitoring of HV transmission and distribution lines and has potential applications to electric-power generating and distribution organizations.

5.6 FO CATV Receiver

Most cable television (CATV) networks deploy copper coaxial cables and hybrid silicon amplifiers consisting of cascaded structures buffered with Darlington-based drivers as output stages. The CATV networks suffer from high transmission losses and high maintenance costs and require high operating voltages. Deployment of an FO network offers several advantages, such as low transmission loss, high gain, flat frequency response, low noise figure, high reliability, built-in immunity against EMI/RF interferences, ability to operate under harsh environments, and excellent compatibility with advance digital transmission techniques [7]. The CATV receiver described here can be used with existing distribution and trunk amplifier systems, thereby yielding greater operational flexibility and improved overall performance with minimum power consumption. The GaAs MMIC technology offers state-of-the art CATV amplifiers with the possibility of integration of surface mount technology, the least expensive materials, and an improved fabrication process. The GaAs MMIC amplifiers provide excellent linearity, improved gain flatness, and low noise figure.

5.6.1 Design Aspects and Performance Parameters

The FO CATV analog receiver, which can be used in the trunk feeder lines of an FO-based CATV network, is essentially a low-distortion, low-noise, wideband receiver with a minimum gain of 30 dB over a wide bandwidth. This analog receiver will boost weak signals with a low return loss and minimum noise figure over an RF bandwidth from 40 to 860 MHz, with output matched to standard 75 ohm cables. The receiver consists of an optical front end with an automatic gain control (AGC) feature and a post-amplifier coupled to a well-matched, wideband power amplifier. The front end of the CATV receiver incorporates an InGaAs/InP photodetector with low input capacitance and negligible gate leakage current. The receiver contains a MESFET-based transimpedance amplifier (TIA). The TIA is best suited for a high-frequency front end, where high sensitivity, low noise, and large dynamic range are the principal requirements. The unique design of a TIA allows the photodetector to operate with near-zero dark current that permits photon conversion with better figure of merit (FOM) over a wide bandwidth. The AGC circuit followed by a post-amplifier keeps the electric power level constant as a function of optical input power over the entire operating bandwidth. The bias circuits are used for the GaAs MMIC assembly consisting of two parallel-connected monolithic amplifiers to maintain desired gain and response over the operating band. The feedback circuit with an open loop gain of 30 dB provides flat gain and low distortion in the band of interest.

5.6.2 Advanced Technologies Incorporated in the CATV Receiver

Several advanced technologies have been incorporated in the receiver design. Deployment of an advanced substrate provides excellent microwave, thermal, and physical properties that are ideal for RF and microwave applications. Heat sink with high thermal conductivity and low thermal resistance is used to remove the heat that is generated instantly from the high-power chips. In addition, during the reflow soldering, the solder is allowed to flow through vials and to make intimate contact with the integrated heat sink at the bottom of the integrated circuits (ICs). This permits parallel routes for heat dissipation from the package to the printed circuit board (PCB) surface and ultimately to the extended heat sink on which the PCB is mounted.

5.6.3 Unique Performance Parameters of the Receiver

The maximum gain of the CATV amplifier is 20 dB with gain flatness between 1.5 and 2.0 dB. Due to the square-law characteristic of MESFET devices and excellent impedance matching of the MMIC amplifier, the composite third-order and second-order cancellations are very impressive. The cross-modulation distortion for the GaAs MMIC receiver is much lower than for silicon hybrid devices. The overall noise figure for the amplifier is less than 2.5 dB. In summary, a monolithic GaAs MMIC-based CATV receiver offers advantages over its hybrid counterpart in terms of cost, reliability, and performance. In addition, massive use of surface-mount devices in the receiver has reduced the fabrication cost by eliminating steps such as chip mounting and wire bonding that are often used in silicon hybrid devices. External transmission line balanced unique networks (BALUNS) are deployed to provide excellent impedance match to 75 ohm cable.

5.6.4 Specific Electrical and Optical Parameters

- Instantaneous operating bandwidth: 40 to 850 MHz
- Input voltage: 12 VDC
- Drain current: 275 mA
- Operating wavelength range: 1300 to 1600 nm
- Return loss: –18 dB
- Receiver gain: 30 dB
- Input and output impedance: 37.5 ohms

5.7 Optical Sensors for Detection of Poisonous Gases and Chemical Agents

Deployment of state-of-the art instruments and optical sensors is necessary for rapid detection of poisonous gases and chemical pollutants. The optical sensors described here are best suited for detection of gas leakage, monitoring of chemical processes, estimation of contamination of metal surfaces, analyzing water quality, and detecting chemical species that contaminate the sample under test. Optical sensors use light as a sensing probe and are nondestructive in nature [8].

Designers of optical sensors deploy the surface plasmon resonance (SPR) technique to detect and monitor poisonous gases, chemical agents, and organic pollutants in drinking water. Optical-sensor performance is dependent on the change in the optical parameters of the sample under test, such as refractive index, film thickness, and mass of the optical film. These parameters are correlated with the change in SPR parameters such as resonance angle, resonance depth, and resonance half-width. Critical elements of the optical sensor illustrated in Figure 5–5 play a key role in the improvement of sensitivity, reliability, and performance level of the sensor.

5.7.1 Surface Plasmon Resonance (SPR) Phenomenon

Surface plasmon (SP) is a surface EM field charge density oscillation that exists at the interface between a metallic surface and a dielectric surface. The magnitude of the EM field is maximized along the surface that decays exponentially on both sides of the interface. The wave factor (k_{SP}) associated with SP is very sensitive to surface roughness, dielectric-coating thickness, exposure of the film to gases, and ambient temperature variation. Excitation of surface plasmon oscillations occurs when the real part of the dielectric $\varepsilon_1(w)$ is negative and the imaginary part $\varepsilon_2(w)$ is negligible [8]. This condition is easily fulfilled in metals. The drastic change in SPR parameters mentioned above can be exploited to design various optical sensors to detect and monitor gas, biochemical agents, and temperature.

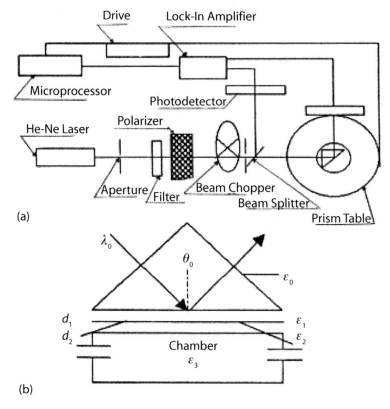

Figure 5–5 *FO sensor for detection of gases and chemical agents: (a) experimental setup and (b) schematic diagram for four-layer system.*

5.7.2 Conditions for Optical Excitation of SP

Optical excitation of SP requires matching of both the energy (optical wavelength) and momentum (wave vector). The X component of the SP wave vector at a given wavelength can be written as

$$k_{SP}^2 = [(k_0^2)(e)\sqrt{\varepsilon_1 \varepsilon_2 / \varepsilon_1 + \varepsilon_2}] \qquad 5.31$$

where k_0 is the free-space wave vector and ε_1 and ε_2 are the complex dielectric constants for medium 1 and medium 2, respectively. Each complex dielectric constant is a function of optical frequency (w) and consists of real and imaginary components. The momentum is too great for excitation from medium 1, as illustrated in Figure 5–5. The modes of the single interface system cannot be excited optically. The method of excitation for SPs is analogous to the prism coupling technique for optical waveguides as shown in Figure 5–5. In the schematic diagram, d_1 indicates the metallic film thickness, ε_1 is the complex relative dielectric constant ($\varepsilon_1 = \varepsilon'_1 + j\varepsilon''_1$), d_2 indicates the coating thickness with complex permittivity of ($\varepsilon_2 = \varepsilon'_2 + j\varepsilon''_2$), and ($\varepsilon_3 = \varepsilon'_3 + j\varepsilon''_3$) is the complex permittivity of the chamber.

5.7.3 Operational Aspects of the Sensor

Successful SPR operation is dependent on sensor-film thickness and the preparation techniques for the metallic and multilayer organic films. For maximum sensitivity, optimum film material and thickness are necessary and the films must be active near ambient temperature. If the optical properties of the film change with time, the sensor performance will degrade. Gold must be selected for the metallic film because it is very inert. The SPR technique yields a potential benefit; namely, the metallic film can be used as an electrode for electrochemical control of sensing various reactions. For organic multilayer films, the films of cross-ether-substituted phthalocynane and cailx resorcinarane can be used because they will respond to nitrogen dioxide at room temperature and they can be deposited on the metallic gold film.

Poisonous gas detection requires placing a thin layer of chemically active film on the metal surface, which SPs propagate. Changes in this layer due to exposure to gas will result in SPR. The parameters that undergo changes due to gas absorption are the refractive index, mass, and film thickness. The refractive index decreases due to swelling of the film and increases due to the addition of gas molecules. In a chemical sensor based on the SPR technique, the presence of chemical agents or species produces a change in the refractive index at the metallic/dielectric interface, which in turn results in a change in the reflected-light intensity or a change in the coupling angle (θ_{SP}) of the SP excitation. By monitoring the resonance condition of the SP, the dielectric constant of the test sample is obtained. If the gas or chemical sample is contaminated by some chemical agents, dielectric constant measurement provides the concentration of the chemical agents in the sample under test. For an SPR-based temperature sensor, the angle of the prism assembly (see Figure 5–5) is set at an optimum value close to 61.5 degrees and the reflected light intensity is monitored as a function of temperature. Reflected-light intensity versus temperature shows a near-linear characteristic over the 20° C-to-60° C temperature range. Threshold changes in light intensity can be seen over this temperature range. This indicates the sensor's sensitivity to even small changes in temperature.

5.7.4 Experimental Procedure for Detection of Gas or Chemical Agent

Successful demonstration of an SPR technique requires an experimental setup (see Figure 5-5) and test procedures. The experimental setup needs a monochromatic light source (He-Ne laser), an aperture followed by a polarizer, a mechanical chopper, and a beam splitter. The optical beams are passed to an LIA. The prism/sample (gold-coated glass slide) assembly is mounted on a motorized rotating table that permits minute changes in the incident angle. The optical signal is detected by a photodiode, and plot graphic software is used to get the reflectivity versus incident angle plot. A dip in this curve occurs at a specific incident angle, which corresponds to the maximum coupling under resonance condition. The SPR curve is characterized by the resonance angle, resonance depth of dip, and resonance half-width. SPR parameters can be calibrated against the optical parameters to design various optical sensors based on the SPR method.

5.8 FO Sensor for Quality Control and Assurance

Automation in manufacturing and quality control/assurance in industrial applications require nondestructive, accurate measurements of angular displacement of various parts in a complex assembly and surface slope of a product with utmost precision [9]. Exceptionally high standards of on-line and in-process quality-control tests need to be performed with emphasis on speed, accuracy, and test sequence in the production environments. Optical sensors are best suited for accurate surface topology and slope measurements to meet stringent quality-control requirements. Optical sensors using optoelectronic converters and matching-signal processing algorithms are integrated in the operating systems that form an integral part of the measurement and characterizing scheme. Measurement accuracy and resolution are dependent on the types of sensors deployed. The FO sensor described here will be best suited for angular and slope measurements on a smooth surface with an accuracy of 3 to 5 microradians.

5.8.1 Operating Principle of the Sensor

The ability of this optical sensor to measure the angular displacement and slope with great accuracy is based on the reflection and scattering of EM waves from the object under investigation. The EM wave incident on the object surface is reflected and partially scattered in different directions depending on the tilt angle and the orientation of the probed area of the surface under measurement. Analysis of the reflected and scattered EM signals from the object surface provides the estimate of angular displacement and slope. Scattered and reflected signals can be determined accurately using either a diffraction model or mirror-facet model. The mirror-facet model is also known as the profile-angle method. In this method, a surface can be approximated by small smooth mirror facets. In the case of the scattered reflection method, the angular-displacement and slope information are provided by the directional orientation of the illuminated mirror facets, and their values can be evaluated in three-dimensional space. The output signals are the intensity-modulated signals. The modulated envelope can be retrieved from the calibrated voltage output of the detector. The signal analysis is based on geometrical optics theory. This sensor can also be used to detect linear displacement, pressure, and even the position of a float. Angular-displacement and slope measurements of an object surface can ultimately be used to determine surface texture with minute details on tilt and orientation of the elementary pixels forming the target or sample surface.

5.8.2 Critical Parameters Affecting the Sensor Performance

The intensity of the reflected light is a function of the illuminated spot shape, coupling to the received fiber, area of the launching fiber, and overlap between the back reflected light and the fiber end face. This light intensity (I) is directly proportional to the square of the fiber core radius (R_c) and to reflected beam shift (x). The reflected beam shift can be written as

$$x = [h \tan(2\delta)] \qquad 5.32$$

where h is the distance between the probe and the object, and δ is the tilt angle between the incident beam and normal to the object. If the tilt or angular displacement (δ) is very small, the above equation can be rewritten as

$$x = [2h\delta] \qquad 5.33$$

When ($x/2$) is less than the core radius (R_c), the intensity can be written as

$$I = C_p[(\pi R_c^2)\{1 - (4h/\pi R_c)\delta\}] \qquad 5.34$$

where C_p is the proportional constant.

The output voltage of the optical sensor is proportional to the incident optical intensity (I) and can be expressed as

$$V = C_{op}[(\pi R_c^2)\{1 - (4h/\pi R_c)\delta\}] \qquad 5.35$$

where C_{op} is the overall proportionality constant.

The above calculations assume that the pixel dimension and their periodicity are large compared to light-source wavelength. The angular displacement and slope are determined by locating the peak of the reflected intensity profile. Since the intensity profile coming out of the launch source is represented by the Gaussian distribution, the far field intensity pattern on the object normal to the fiber will be same. However, for a tilted object, the distribution is nonsymmetric and becomes a non-Gaussian distribution. Under this situation, the peak of the intensity profile will be shifted laterally and is proportional to the angular displacement.

5.8.3 Peak of the Reflected Intensity Profile Under Twofold Rotation of the Facet

The angular displacement of the object in 3-D space involves the rotation of the XY planes around the Y-axis (α), followed by the rotation around the X-axis (β), and vice versa. The peak of the reflected intensity profile can be defined as follows

$$\alpha = [0.5 tan^{-1}(x/h)]$$
$$\beta = [0.5 tan^{-1}(y/h)]$$

5.36

where x indicates the position of the peak along the X-axis, y indicates the position of the peak along the Y-axis, α is the angle of rotation round the Y-axis, and β is the angle of rotation around the X-axis. The position vector indicates the peak amplitude of the reflected intensity profile, which provides the measure of the angular displacement of the test sample. Under tilt conditions, the shape of the reflected spot beam is a deformed ellipse. This means that the reflected and scattered fields sensed by different optical fibers in an array can be interpreted in terms of angular displacement and slope of the test sample.

5.8.4 Experimental Investigation Requirements

An efficient packing structure for the optical sensor is necessary to achieve maximum measurement accuracy. Graded-index optical fibers must be used for these measurements. A graded-index MM fiber using a silica-glass core with a core diameter of 50 microns and cladding diameter of 125 microns is recommended. The selected fiber must have an NA of 0.2, transmission loss less than 2.5 dB/km, and a bandwidth-distance product of 1 GHz-km. The system uses an illumination source of 750 mW and an n.p.n phototransistor as a detector. The power received by each fiber of the array for different angular displacement of the reflecting mirror is recorded. The angular-displacement and slope measurements are obtained by analyzing the direction and magnitude of the reflected beams from the object surface. The data collected from the model can be used to fabricate a compact sensor probe consisting of one beam-emitter fiber and other receiving fibers. The electro-optic (EO) sensor described here is simple and reliable and offers accurate displacement and slope measurement in the range of 0.1 to 0.000001 radians. This measurement technique can be easily implemented under computer control, which has potential industrial applications involving manufacturing of automobiles, trucks, tractors, tanks, and heavy machinery.

5.9 FO Imaging Probes to Study Rock Structures

FO-based Raman imaging probes can be used to study the microscopic structures of rocks or underground objects. According to a professor at The University of California (Los Angeles), morphology does not alone effectively distinguish a fossil from a nonbiogenic mineralogical feature. But by using an advanced imaging technique based on Raman spectroscopy, 2-D chemical maps of a rock sample can be generated. These maps can identify microfossils. The structures that were previously examined, such as an accepten microfossil [10] from 770 million to 3.4 billion years ago, seem to display peak responses at wavelengths ranging from 135 to 160 nm in Raman band. This is characteristic of biogenic carbonaceous material. The chemical maps so generated also fit the optical images of structures, indicating that an optical signal does not arise from the bulk. Research scientists from England and Australia suggest that the geochemical evidence is not conclusive that life was present in the samples investigated. The studies performed by these scientists, who have used Raman spectroscopy, found that the structures display carbon isotopic ratios as an indicator of biogenesis. Digital-image analysis using computer software was performed, which generated montages of the specimens from images at different local depths. However, some scientists question the finding that the morphology of the structures is an indicator of the presence of fossilized microbes. New examination of the site from where the samples were collected suggests that a hydrothermal vent formed the charts at temperatures of up to $350°C$ [10]. A subsequent investigation suggested that it is possible to form hydrocarbons by vents at the above temperatures.

In the Apex charts, researchers indicate that these hydrocarbons, when wrapped around quartz microcrystals, give the appearance of cellular structure. They also suggest examining the alternative possibilities of inorganic synthesis and recommend investigating the findings that isotopically light carbon denotes biogenesis. However, most scientists agree that kerogen, the particular organic matter present in the rocks of which the fossils are composed, is nonbiological in origin and has never been identified in the geological records.

The hydrothermal origin of the charts does not suggest that the microstructures are not microbial remains. Microfossils from the 2.1-billion-year-old Gunflint Formation have been preserved in such an environment, and Archeans are widespread at today's deep-sea vents and hot springs. There is no question that geological maps help address the early sun problems and explain the existence of microfossils from this era. Physicists theorize that the sun is only 70% as luminous as it was 3.5 billion years ago. As a consequence, the sun was an unlikely source of energy for organisms at that time. However, even back then, hydrothermal activity was absolutely necessary for life to exist. Finally, it can be stated categorically that FO Raman imaging sensors can be used to study rock formation and structural details.

5.10 FO Sensors for Site Monitoring

Scientists have suggested that FO sensors employing FBG devices are best suited for monitoring automotive structures and associated components during manufacturing and assembly. The FO system described here is essentially a derivative spectrometer that uses a fiber Fabry-Perot (FFP) tunable filter for wavelength-shift detection and an electronic device for signal processing [11]. Designing advanced, complex automotive structures and heavy vehicles requires comprehensive knowledge of the materials and their properties prior to their use in heavy industrial equipment. The sensor described here allows on-site monitoring of vehicle structures and their critical parts prior to their insertion in the main body. The FO sensor provides strain, pressure, and temperature measurements in the shortest time frame, monitoring when and where a specified part is to be inserted in the structure. Since the FBGs can be easily multiplexed, monitoring systems incorporating such devices can be realized for wide structural applications. Furthermore, when an FBG-based strain sensor is interrogated by an FFP tunable filter, it provides accurate strain measurements under both static and dynamic conditions. This sensor technology offers a cost-effective monitoring system in the manufacturing and assembly of automobiles, trucks, tanks, and heavy machinery.

5.10.1 Operational Aspects of the Monitoring Sensor

An FBG generates a periodic modulation in the refractive-index parameter of a SM optical fiber. Coupling theory states that an energy exchange takes place between the fundamental mode of the fiber and its counterpropagating mode. Therefore, a light wave is reflected back into the fiber under the Bragg condition, and a peak signal is observed in the reflected spectrum with a Bragg wavelength. The Bragg wavelength is given as

$$\lambda_B = [2n_e p] \qquad 5.37$$

where n_e is the effective refractive index of the fundamental fiber mode and p is the grating pitch of the FBG device used by the sensor. When a strain is applied to an FBG device, a shift occurs in the Bragg wavelength (λ_B). The wavelength shift is due to the variation in the grating pitch and in the refractive index of the fiber due to the photoelastic effect. This wavelength shift can be expressed as

$$\Delta \lambda_B = \lambda_B [1 - n^2 / 2\{P_{12} - v(P_{11} - P_{12})\}] \varepsilon \qquad 5.38$$

where P_{12} and P_{11} are optical strain tensor components, v is the Poisson's ratio and ε is the strain applied to the grating. The relationship between the Bragg-wavelength shift and applied strain is nonlinear due to the presence of the Bragg-wavelength term in the above equation. For $\Delta\lambda_B/\lambda_B$ less than or equal to 1%, the above relationship can be considered linear. With good approximation, Equation 5.38 can be rewritten as

$$[\Delta \lambda_B] \approx [(K_\varepsilon) \varepsilon] \qquad 5.39$$

where K_ε is known as the strain-sensitivity parameter, whose typical value is 0.001nm.μ/ε.

A wavelength shift is also observed under temperature variations. This means an FBG device can be used as an FO-strain and/or temperature sensor. But the cross-sensitivity between the strain and temperature must be avoided in sensing applications to preserve high accuracy in the measurements. Therefore, a compensation technique to avoid cross-sensitivity problems must be adopted in field measurements.

5.10.2 Design Concept of the Interrogation System

The reflected spectrum of an FBG can be observed on a spectrum analyzer, which provides accurate measurement of the Bragg-wavelength shift. In the case of mechanical sensing applications, it is necessary to detect vibrations with frequency up to 50 Hz. For this purpose, a derivative spectrometer using a FFP tunable filter along with a fast signal-processing scheme is required to achieve precise strain value under dynamic conditions. The experimental setup to make measurements with a derivative spectrometer sensor is shown in Figure 5–6. The monitoring sensor has three distinct parts: an optical part, optoelectronic part, and an electronic part. These are clearly identified in Figure 5–6.

5.10.2.1 Optical Part Description

This part consists of a low-cost broadband LED optical source operating at 1550 nm with an FWHM of 40 nm. The LED-source output is connected to a 3 dB optical coupler input arm, and the output arm is spliced to the FBG device (top right area), as illustrated in Figure 5–6. The FBG parameters include a Bragg wavelength of 1533 nm, reflectivity of 90%, and FWHM of 0.30 nm. The source light outgoing from the second arm on the input side of the coupler is the reflected spectrum of FBG, which constitutes the input for the optoelectronic part. The FBG is fixed to an automobile steel sample carrying a strain gauge to allow comparative strain measurements. The steel sample is fixed to a vibrating table for characterization under dynamic test conditions.

5.10.2.2 Optoelectronic Part Description

The FFP tunable filter has specific characteristics: a FWHM of 25 GHz, finesse of 200, and a slew rate of 7 V/ms. The filter is driven with a sawtooth electronic signal modulated by a dither signal with a dither rate of 100/sec and an amplitude equal to 1% that of the sawtooth signal, which acts like a reference signal to a an LIA. Thus, only the spectral component of the input signal, which is the electrical output of the photodetector, is amplified at the dither or hoping frequency. The detector output can be expressed in Fourier series, and the spectral components include the FBG spectrum at zero frequency, the first derivative at the dither frequency, the second derivative at twice the dither frequency, and so on. By connecting the oscilloscope to the output of the LIA, it is possible to display the first derivative of the FBG.

5.10.2.3 Electronic Part Description

The interrogation system uses an electronic device for signal processing to detect or measure the strain value. Two digital signals are obtained by differentiating the sawtooth frequency and filtering the LIA output using a Schmitt trigger. A delay detector is programmed to measure the delay between the two digital signals related to the strain. These digital signals are converted into analog signals using an 8-bit digital-to-analog converter (DAC) and are displayed (right side of the electronic section) on the scope. The sampling rate is generally not uniform under dynamic conditions and a simple three-point interpolation algorithm is used to overcome the problem of nonuniformity. Note a sampling rate greater than twice the signal frequency is required under static conditions.

5.10.3 Typical Sensor Parameters

The steel sample is subjected to a vibration test at an initial frequency of 17 Hz with a maximum amplitude of 1600 µε and maximum output voltage of 2.02 V. When the sample is subjected to a vibration frequency of 83 Hz, the maximum strain amplitude is 400 µε when the maximum output voltage is 0.53 V. The performance of the system is compared with that of the electrical strain gauge for vibrations in the same frequency range. The tests indicate that the strain resolution varies between 3 and 10 µε depending on the frequency of vibration. Vibration-signal measurements at a frequency of 50 Hz indicate that test data

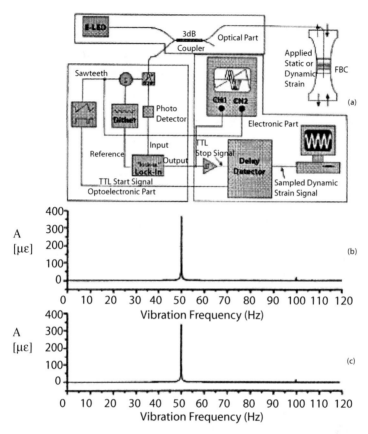

Figure 5–6 *Interrogation sensor using FBG-based strain sensor: (a) experimental setup to measure strain amplitude, (b) strain amplitude from strain gauge, and (c) strain from system.*

obtained with the system described here and the electrical strain gauge are in agreement. In summary, an FBG-based strain interrogating system works well under both the quasistatic and dynamic conditions. Test results obtained using two different methods show good agreement. Temperature compensation can be achieved by using an FBG device loosely attached to the structure under test. Integrating the interrogation system on a small electronic card makes the system suitable for manufacturing of automobiles, trucks, tractors, and heavy equipment.

5.11 Summary

Advanced FO-based sensors are best suited for commercial and industrial applications. Cryogenically cooled optical spectroscopes are applicable for spectral measurements on rare-earth compounds as a function of temperature and magnetic field. Performance capabilities of an AON switch were discussed. This particular switch offers fast switching action, distortion-free transmission, and minimum insertion loss. The switch can be used under severe operating environments due to its robust design. Mathematical expressions for various switch parameters were developed. Benefits for chirped FBG devices were highlighted. An NL pulse switch using nonlinear loop mirrors was also described. This NL pulse switch is free from pulse distortion and is best suited for long-haul transmission systems with data rates exceeding 75 Gbit/s. FO-based optical sensors for precision tracking of high-speed train tracks to avoid accidents were summarized. FO gyros with potential applications in aerospace applications were described, along with the derivation for critical parameters such as coherent length, phase difference, and phase error. Performance parameters of FO sensors for measurement of HV levels in transmission and distribution lines were discussed. Advantages of FO-CATV receivers over their silicon hybrid counterparts were summarized in terms of cost, reliability, and performance. Optical sensors using the SPR phenomenon for detection of poisonous gases, atmospheric pollutants, and chemical agents were described, along with mathematical expressions for light intensity, reflected-beam shift, and peak reflected intensity profiles. FO Raman imaging probes to study microscopic structures of rocks and underground objects were discussed. Advanced imaging technology based on Raman spectroscopy is best suited for 2-D chemical maps of rock samples. A derivative spectrometer using an FFP tuning filter that is useful for wavelength-shift detection was described. This sensor is best suited for in-site monitoring of automobile structures and associated critical parts during manufacturing and assembly.

5.12 References

1. Oreans, P. (1992). A helium cryostat for optical spectroscope using a single-mode fiber optic system. *Cryogenics, 32*(9), 810–812.

2. Perlin, V. E., et al. (2001). Nonlinear pulse switching scheme using cross-phase modulation and fiber Bragg gratings. *IEEE Photonics Technology Letters, 13*(9), 960–962.

3. Boas, G. (2002, May). Laser ultrasound tests train tracks. *Photonics Spectra*, 23–24.

4. Motohara, S., et al. (2000, December 18–20). *Fiber optic gyroscope with single-mode fiber coil.* Presented at PHOTONICS-2000 International Conference, Calcutta, India, pp.30–33.
5. Shupe, D. M. (1980). Thermally-induced nonreciprocity in the fiber optic interferometer. *Applied Optics, 19*(5), 654.
6. Martinez-Leon, L. (2001). Frequency-output fiber optic voltage sensor for high-voltage lines. *IEEE Photonics Technology Letters, 13*(9), 996–998.
7. Johri, S., et al. (2000, December 18–20). *GaAs MMIC fiber optic analog CATV receiver.* Presented at PHOTONICS-2000 International Conference, Calcutta, India, pp. 126–129.
8. Panigrahi, S., et al, (2000, December 18–20). *Optical sensors based on Surface Plasmon Resonance (SPR) measurements.* Presented at PHOTONICS-2000 International Conference, Calcutta, India, pp.726–728.
9. Paul, S., et al. (2000, December 18–20). *Fiber optic sensor for angular displacement and surface slope measurement.* Presented at PHOTONICS-2000 International Conference, Calcutta, India, pp.857–862.
10. Burgess, D. S. (2002, May). Raman imaging probes ancient rocks. *Photonics Spectra,* 22–23.
11. Falciai, R., et al. (2000, December 18–20). *An interrogation system for fiber-Bragg-grating strain sensor for automotive applications.* Presented at PHOTONICS-2000 International Conference, Calcutta, India, pp.471–474.

CHAPTER 6

Fiber Optic-Based Communications and Telecommunications Systems

This chapter describes the performance capabilities and limitations of communication and telecommunication systems utilizing FO components and devices. FO-based optical add/drop multiplexers and demultiplexers are discussed in greater detail with particular emphasis on channel homowavelength cross talk and out-of-band cross talk. FO communication links are described, which are best suited to control the command signals for the transmit/receive (T/R) modules in high-power missile-defense radars. Passive WDM and DWDM technologies involving free-space diffraction gratings, Fabry-Perot tunable passband filters and EDFAs are extensively discussed. Performance capabilities and critical operating parameters of DWDM systems using thin-film filters, fiber Bragg gratings (FBGs), arrayed waveguide gratings (AWGs), and hybrid devices based on free-space optics and diffraction gratings are identified. Advantages and disadvantages of thin-film filters, arrayed waveguide gratings, and free-space diffraction gratings are described with emphasis on channel capacity, channel isolation, and component integration capability.

Performance improvements in long-haul, high-data-capacity transmission lines (4,000 km, 64–40 Gbit/s channels) incorporating distributed feedback (DFB) lasers, counterpropagating Raman amplifiers, and forward error-correction technology are described. Cost-effective networking techniques using WDM and DWDM technologies and incorporating wideband EDFAs and FBGs are briefly discussed. Forward error-correction techniques using sensitive and powerful algorithms are evaluated for long-haul telecommunication systems operating beyond a 5,000 km distance. Performance requirements for wideband Raman amplifiers and EDFAs widely used in long-haul transmission systems are identified with emphasis on gain flatness over a wide spectral bandwidth. Advantages of a forward error-correction technique in Raman amplifiers used by a long-haul DWDM system are described. Applications of AWGs in multiplexing and demultiplexing systems using WDM signals are briefly discussed. Techniques to improve the instantaneous spectral bandwidths of Raman amplifiers and EDFAs are described with emphasis on gain profile over spectral bandwidth exceeding 80 nm.

Techniques to increase network capacity and reduce transmission costs are discussed as a function of channel count, bit error rate (BER), and transmission distance. Performance requirements for optical transceivers with potential applications in a synchronous optical network (SONET), asynchronous transfer mode (ATM), synchronous digital hierarchy (SDH), fiber-channel fast ethernet (FCFI), and gigabit Ethernet are summarized.

6.1 FO Communication Links

FO communication links [1] are best suited for transferring data with maximum security, with minimum insertion loss, and without any RF interference (RFI). An optical link typically involves an LED transmitter, modulator, preamplifier, and photodiode detector (Figure 6–1). The output of the LED transmitter is modulated by the intelligent information to be transmitted and is coupled to an optical fiber transmission line. Modulation of the source is accomplished by modulating the input current. The signal intelligence is recovered through conversion of the optical power, which is then amplified by a preamplifier. The overall insertion loss includes the fiber transmission loss, coupling loss, and detector loss. The data-transmission capability is dependent on the transmitter power output, receiver sensitivity, and link margin to meet specific BER requirements. Data transmission over an optical link can be transmitted with minimum cost, without being susceptible to intentional and unintentional jamming.

6.1.1 Optical Control of T/R Modules in Phased Array Radars

Optical control provides the most cost-effective control of T/R modules in high-power surveillance and missile-defense tracking radars. Recent development of high-speed analog FO links and key circuit technologies offers an effective optical control and interconnects of microwave monolithic integrated circuit (MMIC) in T/R modules deployed in missile-defense tracking radars. Precision amplitude and phase control of RF signals in an electronically steerable phased array antenna are required to meet the stringent sidelobe and beam-pointing accuracy requirements. FO link technology plays a key role in MMIC-based phased array antennas requiring accurate and reliable performance under severe operating environments. In addition, FO technology permits rapid integration of MMIC and photonic components in complex military and space systems [1]. Critical elements of optically controlled phased array antennas using MMIC-based T/R modules are illustrated in Figure 6–2. Deployment of FO control of T/R modules in high-power radars offers multiple-target capability with high tracking accuracy and low probability of false alarm.

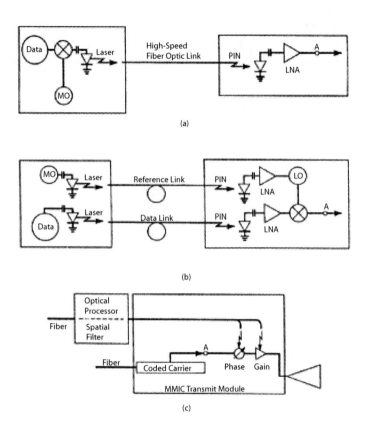

Figure 6–1 *(a, b) High-speed FO link configuration and (c) optical signal processing scheme for the transmit module using MMIC technology.*

6.1.2 Shortingcomings of FO Control

Despite its excellent isolation over a wide bandwidth, high immunity to EMI, electromagnetic pulse (EMP), and rapid-integration capability of optical signal processing, FO links have drawbacks. They require an interface to the MMIC assembly involving an optical transmitter and receiver, which results in additional cost and reduction of link performance in terms of gain-bandwidth product, dynamic range, and signal-to-noise ratio (SNR). However, recent developments have found ways to overcome these problems.

6.1.3 FO Link Configurations and Types

FO links are widely used for transmission of carriers, data, and video signals. Link configurations can be optimized for unidirectional or bidirectional operations. Critical component requirements for these two operations are slightly

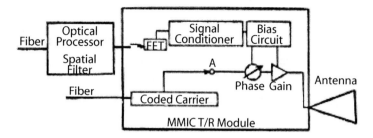

(a) Indirect Optical Control of MMIC-Based T/R Modules

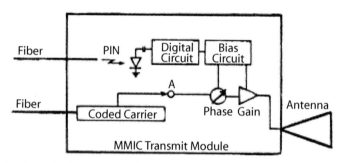

(b) Direct Optical Control of MMIC-Based T/R Modules

Figure 6–2 *Optically controlled MMIC-based T/R modules using (a) indirect method and (b) direct method.*

different. Implementation of WDM techniques in unidirectional and bidirectional FO links involving two signals is possible, as illustrated in Figure 6–3. FO links are capable of sending 40 Gbits/s signals at significantly improved speed and clarity over distances exceeding 600 km without regeneration because of significant improvements in EDFA technology. Using an optical bandwidth of 80 nm, a WDM system has a channel-transmission capacity of 100 WDM channels of 10 Gbits/s, which can yield a transmission data rate of 1 THz over distances greater than 450 km.

FO link configurations are classified according to the coding techniques used by the links. Two types of coding techniques are used: central coding (CC) and remote coding (RC). An FO link configuration using the CC technique consists of a solid-state AlGaAs optical transmitter; wideband optical receiver; and 50-meter, low-loss optical fiber. This type of coding uses resistive matching. An up-converted signal after mixing with a stabilized IMPATT oscillator, is moni-

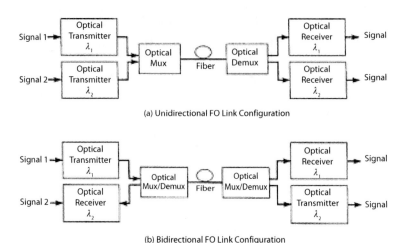

Figure 6–3 *Implementation of WDM technique in (a) unidirectional and (b) bidirectional FO link configuration.*

tored on a spectrum analyzer to detect the quality of the signal. Studies performed by the author indicate that the CC technique cannot be implemented above Ku-band because of excessive losses in the commercially available lasers and detectors at 16 GHz. The RC approach has been successfully demonstrated in 26 and 40 GHz frequency bands by using the nonlinearities of the electro-optic and MM-wave components. Furthermore, the RC approach uses a reactive matched detector circuit that offers better performance over the CC technique, even below 10 GHz operation. In addition, the RC configuration [1] offers a 10 dB improved dynamic range when compared to the CC method because a reactively matched detector circuit is used.

6.1.4 Optical Beamforming

The term "beamforming" applies to the electronic steering of an antenna beam to a specific point defined by azimuth (AZ) and elevation (EL) angles. The adjustment of antenna-beam-shape parameters, such as sidelobe level and 3 dB beamwidth in an electronically steerable phased array antenna is accomplished through the use of phase shifters and gain controllers. The phased array antenna must be reconfigured rapidly, accurately, and efficiently if the phased array radar is to be operated under complex signal environments. This places severe demands on the power and speed of the computing system and requires a parallel optical signal-processing scheme.

In an electronically steerable phased array antenna, the excitation to each antenna element must be controlled to achieve amplitude tapering to meet sidelobe-level and AZ and EL steering angle requirements. In the case of optically controlled phased array antenna, the phase and amplitude values for each radiating element are a function of optical-signal amplitude. Two beamforming techniques can be distinguished depending on whether the phase shift and/or attenuation is implemented in a microwave or an optical domain.

6.1.4.1 Beamforming in Microwave Domain

Beamforming in a microwave domain uses microwave phase shifters, attenuators, and other RF devices in situ of MMIC T/R module. The control devices are actuated by the control signals produced by GaAs digital circuits and transmitted via optical fibers. The optical fiber link transfers the beam control signals from the CPU to the T/R modules where they are detected, decoded, and amplified to actuate the conventional phase-shift and gain-control circuits. The control wires are replaced by the optical fibers. A monolithically integrated GaAs optical interface chip is used to detect 1 Gbit/s multiplexed control signals. High phase accuracy requires a BER value better than 10^{-9}. Photodetectors coupled with preamplifiers and buffer amplifiers with high isolation are needed to maintain stable antenna performance. These components are interfaced with a digitally controlled 5-bit phase shifter and a RF power amplifier in a T/R module to achieve optimum RF performance.

6.1.4.2 Beamforming in Optical Domain

Beamforming in an optical domain requires FO networks to distribute control signals in conjunction with the parallel optical processing scheme illustrated. As stated earlier, in a T/R module the excitation to each antenna element is a function of phase to steer. The antenna beam is given direction by specified AZ and EL angles and amplifier gain to provide amplitude tapering to meet sidelobe requirements [2]. In this approach, amplifier gain and phase values needed for each T/R module are a function of LED-source intensity. The antenna beam patterns in both planes can be optimized using variable optical spatial filters. Optimized antenna beam patterns require optimum values of optical gain and phase angle. Both the phase-shift and gain-control, as a function of optical-source intensity, must provide linear response over the operating RF bandwidth to achieve the optimum antenna performance. Beamforming with FO links is based on true-time delay phase shifts provided by the FO lines connecting the CPU with the T/R modules, as illustrated in Figure 6–1. The true-time delays can be achieved either by changing the effective length of the optical fiber between the CPU and the T/R module involved or by varying the group velocity of the optical signal carrying the microwave information on the FO link. Changing the path length on the FO link can be accomplished by switching the length of fibers. The distribution of high-frequency and high-data-rate signals requires several narrow microwave beams to track multiple targets. Multiple-target tracking capability of radar in battlefield environments requires beam-forming in an optical domain [2].

6.2 Optical Communications and Telecommunications Systems

The market for FO-based communications, telecommunications, and long-haul (LH) transmission networks using SONET/SDH technologies is now trying to correct itself after declining since 2000. One FO market showing promise is the metro-networks SONET/SDH. However, this market faces challenges from Ethernet. Ten-gigabit Ethernet (10 GE) is in great demand and some of the architectural alternatives include Ethernet over SONET, switched Ethernet over fiber, and Ethernet over WDM. The business outlook looks strong for optical transducer modules, optical switches, and routers. The latest optical module driver is designed for SONET/SDH OC-192 or STM-64. In addition, 10 GE systems have been developed for long-haul transmission, point-to-point (PTP) systems, and DWDM applications involving different transmission lengths, as shown in Figure 6–3.

6.2.1 DWDM Systems

The market for DWDM systems has grown at a staggering pace. As more vendors are entering FO-based businesses, wavelength-handling capacity is growing exponentially. A market survey indicates that the number of carriers deploying DWDM systems has grown faster than the number of vendors selling them [3]. For the first four years of DWDM development, products focused almost exclusively on long-haul transmission that typically represented unregenerated spans of up to 600 km. The development of Raman amplifiers, along with forward error-correction techniques, has stretched the distance for the DWDM transmission without regeneration. There is a great demand worldwide for the deployment of DWDM technology in ultralong-haul communications systems. Current trends indicate that DWDM technology is migrating to the end-user, into access rings, enterprises, buildings, and homes as illustrated in Figure 6–4.

6.2.2 Techniques to Increase Channel Capacity and Count

Optical-communications equipment suppliers are expecting increased wavelength capacity and high channel-count capacity. This capacity is doubling every year. For 2.5 Gbit/s transmissions, channel-count capacity has progressively increased from 8 to 16, 32, 40, 64, 80, 96, to 320 channels. In addition, manufacturers have increased channel counts in several ways, including the use of narrow channel spacing on the filter. Filters with channel spacing of

6.2 Optical Communications and Telecommunications Systems

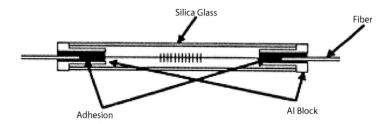

(a) Bragg Grating Architecture for Dispersion Compensation

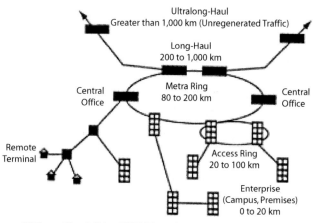

(b) Current Trends Using DWDM Transmission Lines

Figure 6–4 *(a) Bragg grating architecture trends for dispersion compensation and (b) the current trends for DWDM transmission lines.*

200 GHz are currently used, but such narrow channel spacing can make the system more expensive. A cost-effective approach is to take wider-spaced filters and use an interleaver to achieve narrow channel spacing, thereby reducing the channel spacing to half of the output wavelengths and doubling the channel counts. The latest technique to increase the channel counts involves the use of wideband EDFAs currently operating in C-band (1530–1565 nm range). But L-band amplifiers operating over the 1570-to-1610 nm spectral region can double the number of channel carriers capable of transmitting over a single optical fiber with a length of 1000 km (see Figure 6–4) and meeting various end-user requirements.

6.2.3 DWDM Application Requirements

DWDM systems are best suited for LH telecommunications and metro-communications. Multiplexing and demultiplexing hardware are used by these systems, but their costs can be offset over long spans using wideband optical amplifiers capable of amplifying all wavelengths simultaneously. For a typical LH system, one can expect a number of minor nodes between a pair of major switching nodes. Equipment at the switching nodes aggregates the signal quality. To avoid this problem, several separate time-division multiplexed (TDM) signals at 2.5 or 10 Gbit/s are used, each of which occupies its own optical channel. A WDM system can pack many optical channels into a single fiber.

In a PTP system, optical signals are directed to the switching node on the other end. In systems with intermediate minor nodes, the primary switching node aggregates multiple signals into one or more channels. Those signals destined for the same node are directed into the same wavelength channel. The add/drop multiplexer at each intermediate node drops off the channel with locally routed signals. The terminal-switching node at the end then reaggregates signals according to their destination. Standard LH systems span distances of several hundred kilometers between the terminal switching nodes, with optical amplifiers included to boost signal strength. Terminal nodes convert signals reaching them from optical to electronic form and process them for further transmission. Optical channels may be regenerated for transmission to another node and finally for distribution to different destinations. Traditional telephone companies are common carriers, which transmit signals for any end user who wants the service. Some carriers install high-speed systems to lease capability to other companies that carry signals for their end users. These companies have different requirements with fewer switches and longer transmission distances. To meet such requirements, improved DWDM systems are being developed.

6.2.4 Unique Capabilities of DWDM systems

Studies performed by the author and other communication engineers indicate that a DWDM system can transmit signals up to a few thousand kilometers without regeneration [4]. DWDM systems may be installed as upgrades to existing WDM systems. Many carriers are incorporating DWDM equipment to the existing fiber to gain extra bandwidth with minimum cost. Bolting new DWDM transmitters onto old optical fiber with high dispersion poses an operational problem. This problem can be overcome by incorporating new optical amplifiers with multi-channel capabilities. This approach is cost effective and quicker because it is easier to upgrade an existing fiber than to install a completely new system.

6.2.4.1 Critical Design Aspects of DWDM System

Design of an LH DWDM system requires comprehensive trade-off studies involving transmission distance, spacing between optical channels, data-rate requirements, dispersion compensation, dispersion slope, optical amplifier gain and spacing, total span between regeneration points, equalization of gain over optical channels, and add/drop multiplexing requirements. The trade-off between channel spacing and the maximum TDM data rate per channel is a fundamental one for all DWDM systems. Channel spacing is set on a frequency grid, with the optical carrier signals at intervals of 50, 100, or 200 GHz for a DWDM system. A carrier's signals must be spaced widely enough to keep the optical channels from overlapping and to allow demultiplexing optics to separate them at the receiver end. Insufficient separation will cause cross talk between the adjacent channels, thereby degrading the DWDM system performance.

The optical bandwidth required for each channel is contingent on the modulation format, which is simple amplitude modulation used by the current systems. This means that the 2.5 Gbit/s optical channels can be separated by 50 GHz, but 10 Gbit/s channels require 100 GHz separation from adjacent channels going in the same direction. With bidirectional transmission, 10 Gbit/s signals in opposite directions can be interleaved 50 GHz apart, so that the next channel going in the same direction is separated by 100 GHz. This approach can reduce the cross talk to an acceptable level. Cumulative dispersion does not affect channel-separation requirements because it causes signals to spread out in time rather than in wavelength. Pulse spreading caused by dispersion does limit the maximum data rate that can be transmitted through optical fiber.

6.2.4.2 Impact of Dispersion on DWDM System Performance

Fiber dispersion complicates the design of an LH DWDM system. Dispersion compensation is relatively simple at a single operating wavelength. A typical solution to this problem requires the addition of a compensating fiber with chromatic dispersion opposite in sign and roughly equal in magnitude so that the overall dispersion over the entire span is negligible. Dispersion compensation becomes more difficult in DWDM systems operating at multiple wavelengths because chromatic dispersion is dependent on the operating wavelength. Dispersion-compensation techniques involving Bragg grating devices (see Figure 6–4) can reduce dispersion to acceptable levels over the entire range of wavelengths being used for transmission. Depending on the use of fiber and compensation technique, this can limit the range of wavelengths available for DWDM transmission or limit the data rates for some wavelength channels. Total dispersion is dependent on transmission distance and type of compensation device used. Optical transmitters and receivers do not require dispersion compensation for a 100 km transmission distance, do require some compensation for a

300 km span, and do require an elaborate compensation technique for a span of 800 km or more.

6.2.4.3 Impact of Various Effects on DWDM System Performance

Nonlinear effects such as four-wave mixing (FWM) can affect DWDM system performance. Some complications can come from the fact that LH DWDM systems are not simple PTP optical links. Degradation in system performance can occur if optical channels from intermediate points inserted at add/drop multiplexers are not balanced in strength with other optical channels in the same fiber. A span employing different types of fiber and optical cable with different fiber counts can affect system performance.

Optical networking adds additional complications in system performance. Existing designs assume the systems terminate at the two end points, with optical signals converted to electrical format and reorganized at the terminal nodes. With optical networking, DWDM channels will pass through terminal nodes to another portion of the network. With current systems, this will require regeneration to clean up the signal sufficiently for further transmission.

6.2.5 Techniques to Stretch Transmission Distances

New technologies are required to stretch the transmission distances or spans to greater than 2000 km for long-haul optical systems. Three distinct technologies integrated with careful dispersion management are available for this purpose. These technologies include distributed raman amplification (DRA), return-to-zero (RZ) pulse transmission, and forward error correction (FEC) code technique. All these technologies allow optical amplifier spacing of 100 km in very long systems. DRA can be distributed in part of the transmitting fiber, with the pump beam propagating in a direction opposite to the signal. A smooth gain is possible with an EDFA. Note RZ coding requires faster modulation than conventional non-RZ coding because the signal is required to drop back to zero at the end of every bit interval, thereby producing a more robust signal. FEC requires additional bits into the signal so that the receiver can spot and correct errors, if any. Special circuits are used in the receiver to process the optical signals and check for errors using more powerful and sensitive algorithms. This technique improves the receiver sensitivity and reduces the BER. Refinements in optical systems with higher channel counts and longer transmission distances are sure to continue.

6.2.6 DWDM System Performance Limitations from Various Sources

LH DWDM optical systems always contain chains of analog optical amplifiers, which are subject to their own performance limitations. Every optical amplifier introduces a certain amount of spontaneous-emission noise across the range of operating wavelengths. Furthermore, successive amplifiers in the chain amplify both the signal and noise. However, reducing the gain per amplifier stage and adding more amplifiers will reduce the total amount of noise that accumulates over a fixed distance. In addition, the gain variation of EDFAs as a function of wavelength over the operating spectral range will also affect the LH DWDM system performance.

A system consisting of ten amplifiers, each with 3 dB gain difference between channels, would accumulate a differential gain of 30 dB. This degrades transmission of the weakest channels as the strongest channels collect most of the optical energy. The higher the input power, the more uniform gain (i.e., less gain variation) becomes across the spectral range. Another effective method for lower gain variation is to split the gain spectrum of the EDFA into two regions: the C-band (1530 to 1635 nm) and L-band (1570 to 1620 nm). Another proven technique involving optical filters that can smooth out the differential gain across the EDFA band is achieved by attenuating stronger wavelengths more than the weaker ones. It is interesting to mention that a Raman amplifier can boost weaker wavelengths or transfer power from stronger optical channels, thereby making the total gain spectrum more uniform.

Increasing the number of channels puts heavy demands on the laser sources that pump optical amplifiers. The more channels, the more pumping is required. Furthermore, you simply cannot add more channels at the transmitter and receiver end because that will unquestionably increase the system cost and complexity. However, it is imperative that amplifiers have adequate pump power to support the extra channels.

6.2.7 Impact of Nonlinear Effects on Data Rate

SM optical fiber is desirable for carrying a high data rate over a LH FO optical link. The growing bandwidth demand can be satisfied using a DWDM approach with low-dispersion fibers. The DWDM approach enhances the effective data rate on a fiber link by increasing the number of wavelength channels with the spectral range. However, the optical power level in the fiber increases proportionately with the number of channels since each channel has to carry a minimum amount of power so that it can be detected at the receiver with a certain SNR. But the increase in power level can lead to nonlinear effects, such as FWM and cross-phase modulation (XPM) between the copropagating channels, which will degrade the system performance. The dominant nonlinearity in an optical fiber is the Kerr-type nonlinearity [5]. This is defined by a nonlinearity-based refractive index, which is written as

$$n_{NL} = [(3P\eta_0)(O_{fun})]/[8n^2 A_{eff}]$$ 6.1

where P is the optical power in the fiber, η_0 is free space impedance, O_{fun} is the objective function, n is the core refractive index, and A_{eff} is the core effective area. Note the objective function has to be maximized with respect to fiber-profile parameters. Thus, for a given optical power, the nonlinear coefficient is inversely proportional to the core effective area. In other words, reduction in the FWM nonlinear effect requires an increase in the core effective area. The core effective area is dependent on the modal field distribution within the fiber. One of the most dominant nonlinear effects in a DWDM system is four way mixing (FWM). Due to FWM nonlinearities, the copropagating multiple signal channel mixes with each other to generate a sum and difference signal channel, which can sap power from and overlap with the original signal channel, leading to degradation in system performance. High FWM mixing efficiency requires low phase mismatch, which is proportional to the dispersion and dispersion slope of the fiber. The FWM nonlinear effect is more serious in WDM communications systems compared to DWDM systems.

In a high-speed optical communication link, dispersion is an important parameter, in addition to the core effective area, because it affects the nonlinear characteristics of an optical fiber. The optical fiber should have the lowest possible dispersion and the highest possible core effective area over the operating spectral range. Based on foregoing statements, a depressed core refractive index profile for a large-effective-area, non-zero-dispersion fiber is best suited for DWDM systems. Furthermore, both the core effective area and dispersion characteristics can be optimized by the objective function using the profile parameters. The optimum refractive-index profile can yield a core effective area of 110 microns^2. For the optimum refractive-index profile, the dispersion varies linearly from 2.5ps/nm-km to 4.5 ps/nm-km, with a dispersion slope of 0.065 ps/nm^2-km over the 1530 to 1560 nm spectral range. To meet the performance parameters stated above, the bending loss for the fiber must be less than 0.003 dB/m for a bend radius of 10 cm or 0.1 m.

6.3 Performance Requirements for WDM Systems

Performance requirements for WDM-system applications are slightly different from those for DWDM applications. Custom-tailored optical fibers are required for high-performance WDM systems. In addition, adjustment of chromatic dispersion, dispersion slope, and fiber-core effective area is critical. A 1310 nm WDM communication system using standard SM, custom-tailored optical fiber is widely deployed even now due to its high reliability and superb performance [6]. SM fibers used in systems have an impressive low attenuation of 0.2 to 0.3 dB/km at a wavelength of 1550 nm and a dispersion close to 17ps/nm-km. This dispersion affects the performance of a high-speed transmission system. Research studies performed on chromatic dispersion indicate that dispersion can be improved by varying the refractive-index profile of the fiber. Even the zero-dispersion-shifted optical fibers are not suitable for WDM systems because they allow nonlinear effects to build up.

6.3.1 Impact of Dispersion Control on High-Speed Transmission Systems

As stated earlier, large dispersion can limit transmission speed, whereas a little dispersion is necessary to combat the FWM problem. A nonlinear effect arises when three channels are combined to generate a weak signal at a fourth optical frequency. Since the channels are spaced uniformly, the fourth optical frequency interferes with another channel and generates optical noise. With low dispersion, the signals stay in phase over long distances, which results in FWM. Furthermore, a little dispersion spreads the signals, thereby reducing their interaction and damping FWM. Too much dispersion limits transmission span and requires extensive dispersion compensation.

A further complication comes from the slope of the dispersion curve and the broad operating bandwidth of WDM systems (see Figure 6-5). For a standard SM fiber, the typical slope is about 0.08 ps/nm^2-km at 1550 nm wavelength. If the same fiber has to carry signals over a 100 nm spectral bandwidth, the dispersion may vary from 2 to 10 ps/nm-km. This means that lower-dispersion operating wavelengths can go five times further than the higher-dispersion wavelengths without any dispersion compensation.

Sending many wavelength channels through the tiny core of an SM fiber produces very high power density, thereby making them vulnerable to nonlinear effects. High input power levels are desirable for long-distance transmission, but high power density is not acceptable. Using large-size optical fibers can reduce high power density. An SM fiber transmits light in a region that extends beyond the core, defined by the "mode-field diameter" or effective core area. Increasing the mode-field diameter from the typical diameter of 8.4 microns to 10.8 microns nearly doubles the effective area from 55 to 100 microns^2. This reduces the power density by a factor of two and consequently minimizes the performance degradation from nonlinear effects.

6.3.2 Optimization of Fiber Design Parameters

Optimization of fiber design parameters must be given serious consideration if reliable and improved performance is desired from WDM or DWDM systems. Extreme material purity is critical for the fiber design, except for eliminating the 1380 nm water absorption peak. Fiber developers have demonstrated more care and flexibility in controlling the dispersion and effective core area, both of which are dependent on the refractive-index profile or the variation in the refractive index as a function of distance from the center of the core and mode-field diameter.

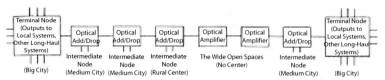

(a) DWDM System Signals between Terminal Switching Nodes, with Add/Drops and Optical Amplifiers at Intermediate Points.

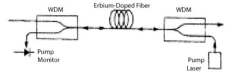

(b) EDFA Configuration for WDM Systems

Figure 6–5 *(a) DWDM system and (b) WDM system configurations using EDFAs at intermediate points.*

In the case of a photosensitive-cladding fiber, the core material is germanium-doped silica glass, like standard fibers. The inner cladding is silica glass co-doped with germanium and fluorine, which decreases the refractive index of the silica glass. The outer cladding is pure silica glass. This photosensitive-cladding fiber enables the formation of a refractive grating in the area larger than the mode-field diameter. This leads to reduction of the change of the mode field between the ultraviolet (UV)-exposed and -nonexposed regions and makes the mode field very stable in the grating region. Thus, by changing the fluorine concentration in the inner cladding, one can control the refractive-index difference between the core and the inner cladding layer. Similarly, other options are available to optimize the refractive-index profile, the dispersion, and the dispersion slope of the SM fiber.

6.3.3 Application of FBGs in WDM Systems

FBGs are widely used as gain-flattening filters for WDM and DWDM systems, pump-reflection filters, emission-rejection filters for EDFAs, chromatic-dispersion compensators, and sensor devices. The most critical element of the system is the temperature-compensation mechanism, as illustrated in Figure 6–6. Temperature compensation is accomplished by inducing a strain with negative temperature dependence, which can mitigate the temperature dependence of the Bragg wavelength. The temperature dependence of the Bragg wavelength in an FBG is close to 1 nm/100°C. The mechanism in the figure is acceptable for a WDM system, but it is too large for DWDM-system applications. To compensate for this temperature dependence, a specially designed package is required, which is displayed in Figure 6–4.

The temperature dependence of the Bragg wavelength for an uncompensated optical FBG is given as

$$[d\lambda_B / dT] = (2 p_g)[dn/dT + n\alpha_{tec}] \quad 6.2$$

where T is the temperature, λ_B is the Bragg wavelength, p_g is the grating period, n is the refractive index, and α_{tec} is the thermal expansion coefficient.

A new term, $(d\varepsilon / dT)$, with a negative value of about –0.7 is introduced in Equation 6.2, and the new expression for the temperature dependence of the Bragg wavelength can be written as

$$[d\lambda_B / dT] = (2 p_g)[dn/dT + n\alpha_{tec} + d\varepsilon/dT] \quad 6.3$$

where ε is a longitudinal strain in the fiber employed.

An aluminum block at each end of a silica glass tube incorporating an FBG is required (see Figure 6–4) to produce the longitudinal strain. The thermal expansion coefficient for the aluminum block is $2.5 \times 10^{-5}/°C$ and for the silica glass tube is $0.055 \times 10^{-5}/°C$. The difference, which is close to 50:1, induces a strain with negative temperature dependence on the FBG. By adjusting the ratio of aluminum block length to the silica glass tube length, one can control the temperature dependence of the Bragg wavelength (B). Using this technique, a temperature dependence of less than 0.5 pm/°C can be achieved over the temperature range of –45°C to 85°C for a temperature-compensated FBG. This boils down to a temperature dependence of less than 0.05 pm per 100°C, which is stable enough for use in DWDM transmission systems. The temperature-compensated fiber Bragg grating devices are used to provide various functions, as illustrated in Figure 6–6.

6.3.4 Techniques to Boost Bandwidth of WDM Systems

The explosive growth of Internet traffic and high data-transmission capacity has placed severe demands on the operating bandwidth of WDM systems. The current communications systems deploy the "low-loss" windows in the optical fiber around 1300 and 1500 nm. The 1500 nm system takes full advantage of the matured EDFA technology in the 1530-to-1565 nm conventional IR band, known as the C-band. Note the bandwidth demand increases by a factor of two every two years, which means that the 35 nm window will not be sufficient to meet future Internet traffic requirements. The stage is set for new optical amplifiers and fiber

172   Chapter 6: Fiber Optic-Based Communications and Telecommunications Systems

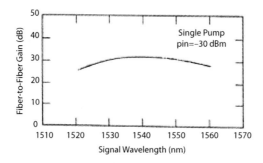

Figure 6–6   (a) Critical elements of an EDFA and (b) most desirable response of the amplifier over the band.

technologies to expand the current usable bandwidth of FO networks at least by a factor of ten to meet future requirements.

6.3.4.1 Optical-Amplifier Requirements to Boost Operating Bandwidth

In an optical amplifier, energy is inserted into a gain medium that consists of molecules, atoms, or electron-hole pairs. The added energy excites the gain medium to higher energy levels. When a photon from an incoming signal passes through the gain medium, that photon can stimulate the emission of an identical photon, leading to amplification of the incoming signal. Since a solid-state amplifier (SOA) suffers from polarization sensitivity, coupling inefficiency, and intermodulation distortion, other techniques to boost the amplifier bandwidth will be discussed.

The gain in an EDFA is the result of stimulated emission of photons from an excited population of Er^+ ions within the optical fiber. This means that concentration of Er^+ ions in the fiber will determine the increase in the amplifier bandwidth. The usable gain of an EDFA falls between 1530 and 1560 nm, which is known as the C-band window. EDFA requires other components such as pumping lasers, isolators, pump circuit configurations, gain-flattening filters, optical couplers, and emission rejection filters, as shown in Figure 6–6. However, the EDFA bandwidth is to some extent affected by the performance parameters of these components.

In recent years, significant research and development activities on C-band EDFAs has focused on techniques to flatten gain over a wide spectral region, to increase power-handling capability, to reduce the noise figure, and to improve the DWDM channel capacity. Because of a rapidly growing demand for bandwidth, a major push today is for the development of amplifiers in the longer-wavelength band from 1563 to 1610 nm (L-band), as illustrated in Figure 6–7. The intrinsically lower gain presented by the L-band EDFA poses a problem. However, by deploying longer lengths of the fiber and band-select filters to reject C-band signals, this problem has been overcome. But some other problems are popping up with the longer-wavelength amplifiers, such as high insertion loss due to longer lengths of fiber and amplified spontaneous emission. These problems can be overcome by operating away from the zero-dispersion wavelength of the fiber.

In the case of LH communications systems or long-distance metropolitan networks, chromatic-dispersion $D(\lambda)$ and dispersion-slope $dD(\lambda)/d\lambda$ characteristics of the fibers must be given serious consideration. The chromatic dispersion is defined as

$$D(\lambda) = (S_0/4)[\lambda_0^4/\lambda^3] \, ps/nm\text{-}km \qquad 6.4$$

where S_0 is the zero-dispersion slope whose value is less than 0.092 ps/nm^2-km, λ_0 is the zero-dispersion wavelength (nm) that varies from 1301.5 to 1321.5 nm, and λ is the operating wavelength.

Computed values of dispersion as a function of zero-dispersion wavelengths and operating wavelengths are summarized in Table 6–1.

The latest research indicates that new fiber technology, such as all-wave fiber (AWF) developed by Lucent Technologies and large effective-area fiber (LEAF) developed by Corning Inc., permits rapid expansion into other wavelength regions. AWF eliminates the absorption of the water peak between 1350 and 1450 nm, whereas the LEAF allows transmission of higher powers with minimum loss and smaller dispersion. Each of these fibers offers different technologies, leading to a significant boost in the optical-amplifier bandwidth. Integration of these technologies in EDFAs or Raman amplifiers opens the entire

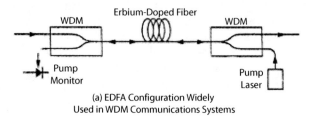

(a) EDFA Configuration Widely Used in WDM Communications Systems

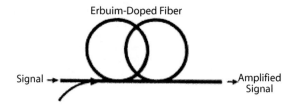

(b) Simple Block Diagram of the EDFA

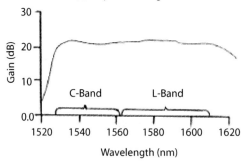

(c) Typical Gain Variations for the Two Spectral Bands

Figure 6–7 *(a) EDFA configuration, (b) simple block diagram of the EDFA, and (c) typical gain variations for two bands.*

spectral range from 1250 to 1625 nm for use in communications systems, which will expand the current 35 nm window to 345 nm.

Most SM fiber used in communication networks is dispersion-shifted fiber designed for use at 1550 nm wavelength. By making the zero-dispersion wavelength at 1550 nm (that is, non-zero dispersion shifted), undesirable nonlinear effects such as FWM, self-phase modulation, and stimulated Brilloun scattering are generated. These nonlinear effects can be overcome by moving over the L-band region. Successful development of L-band EFDAs will find immediate applications in DWDM systems. L-band amplifier capability can be added to an

Table 6–1 Computed values of dispersion as function of zero-dispersion wavelength (ps/nm-km).

Wavelength (nm)	λ_0 (1310 nm)	λ_0 (1321 nm)
1500	0.201	0.208
1550	0.180	0.188
1600	0.165	0.0.171
1650	0.151	0.156

existing C-band amplifier using FO combiners and band splitters so that certain bands only go to the appropriate amplifier.

6.3.4.2 Alternate Techniques to Boost Amplifier Bandwidth

Optical signals can be amplified using the stimulated Raman scattering (SRS) concept, which actually came before the birth of EDFA. SRS is a physical phenomenon in which the pump photon is converted into an optical photon and a lower-frequency photon. When the pump power is high enough, the incoming signal stimulates the process and provides signal amplification over wider bandwidth than that available currently from EDFAs. Development of Raman amplifiers had been minimal because they require high pump power levels. Now, however, the fundamental power problems are solved due to the availability of high-power solid-state lasers. Expanding the total bandwidth of fibers is necessary, which requires combination of many technologies. In addition, potential fiber dopants such as Er^{3+}, thulium (Tm) and ytterbium (Yb) are being explored to boost the fiber bandwidth without impacting its dispersion characteristics. The hybrid amplifier design concept that combines two technologies shows great promise for bandwidth improvement.

6.4 Capabilities of Metropolitan WDM Systems

Optical networking technology is the most critical requirement for metropolitan WDM systems. As the metropolitan WDM technology becomes more cost-effective in the near future and bandwidth demand continues its exponential growth, widespread deployment of WDM technology is inevitable. This technological evolution will drive carriers to expect new solutions, that will permit maximum use of existing networks combined with technical and economical advantages in next-generation metropolitan WDM systems.

One of the biggest advantages is expected in the area of all-optical networks that will offer several key operational benefits, including the ability to manage wavelengths—not optical fibers—which will lead to significant increases in revenue for service providers. Additional benefits include less expensive installation, maintenance, and provisioning. Increased path length and dispersion-management issues need to be addressed in the design of future metropolitan networks (see Figure 6–4). New optical fibers must be optimized for metropolitan-area networks of lengths from 100 to 300 km. Future architectures for emerging metropolitan-based networks require optimization, especially for the 1550 nm window, to be more cost effective. Cost-effective operation can be achieved by using matured component technology, by minimizing the dispersion limitations of these components, and by integrating the EDFA technology in the 1550 nm spectral window.

Silica SM fiber (SSMF) was the fiber of choice for applications in metropolitan networks. This fiber is characterized by a zero-dispersion wavelength close to 1310 nm and possesses a positive profile across the EDFA window in the vicinity of this wavelength. This is in contrast to the MetroCor fiber, in which the zero-dispersion is located at the opposite end of the window close to 1640 nm. The dispersion of the optimized fiber is negative across the EDFA operating in this spectral region. However, the MetroCor fiber was developed by Corning Glass, Inc. especially for the amplifiers operating in the window between 1530 and 1625 nm. Note it is the negative dispersion profile that offsets the positive chirp characterizing the direct modulated laser (DML) source, which ultimately leads to pulse compression at the beginning of the signal path. It is the pulse compression that leads to additional uncompensated path lengths ranging from 200 to 300 km. This operational benefit makes the MetroCor fiber most suited for metropolitan DWDM networks, especially those networks based on optical transparency.

Based on future-growth projections, the next generation of metropolitan systems will most likely consist of densely spaced (100 to 200 GHz) wavelengths in the conventional and long-wavelength EDFA bands (1530 to 1620 nm) at 2.5 Gbit/s transmission rates. Negative-dispersion-shifted fibers (–DNZDSF) have the inherent capability of providing even greater dispersion tolerance (up to four times compared to +DNZDSF) due to the interaction with the positive chirp characterizing the direct modulated optical sources. The performance of SSMF and DNZDSF fibers is expressed in terms of Q (dB). For an SSMF fiber, a Q of 6.5 dB represents the worst-case performance, whereas the worst-case performance for a MetroCor fiber occurs at a Q of 11 dB. The impairment in system performance is due to the dispersion limitations of an SSMF fiber. If dispersion compensation is incorporated into the network, then a more realistic performance comparison between the two fibers can be achieved. However, incorporation of a dispersion-compensation device in the metropolitan environment is not only costly but also very complicated. Regardless of the fiber types used in the metropolitan networks, cost is the most important consideration in the system design.

6.5 Multiplexing and Demultiplexing Techniques

A reconfigurable network provides flexibility and wide bandwidth, which permits carriers to realize a significant return on their investment in infrastructure. The significant demands that a reconfigurable network places on DWDM components have led to a variety of sophisticated multiplexing and demultiplexing techniques, as shown in Figure 6–8. Current DWDM techniques can deploy FBGs, thin-film filters, or hybrid devices based on free-space optics and diffraction grating. The application of the above technologies is governed by factors such as price per channel, performance, footprint, scalability, manufacturability, power consumption, and channel counts [6].

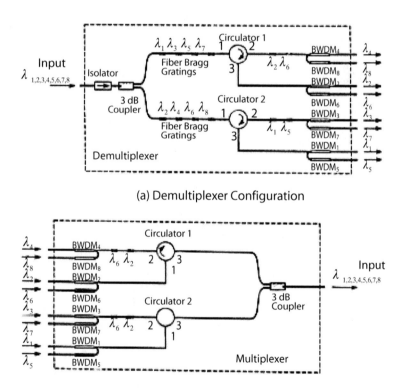

(a) Demultiplexer Configuration

(b) Multiplexer Configuration

Figure 6–8 *Typical (a) demultiplexer and (b) multiplexer configurations widely used in WDM and DWDM communications systems. Courtesy of Johns Wiley and Sons, Inc., New York.*

6.5.1 High-Channel-Count Techniques and Applications

Major networks run many multiplexing and demultiplexing channels through single optical fibers to achieve cost-effective operation. Typically, 40 channels for metropolitan hubs and 80 or more for long-haul hubs are deployed. This requires a technology capable of providing high channel count and excellent uniformity. The most popular approach uses thin-film filters coupled by circulators or interleavers to achieve high channel counts. Since the efficiency of these devices is not dependent on thermal compensation and their passband response is relatively flat, layering of many filters in a system is feasible. The above approach suffers from some manageable problems. Although single thin-film filters have excellent optical properties, their cumulative losses can degrade the overall DWDM system performance at high channel counts when cascaded in series with other thin-film filters, circulators, and interleavers. The prices of these devices have dropped significantly recently; nevertheless, the economics of cascaded thin-film filters are poor because the interleavers, circulators, and other coupling devices must be included in the overall system procurement cost. In addition, DWDM systems deploying cascaded thin-film filters are bulky, often requiring 19-inch racks.

An alternate approach for high-channel-count DWDM filters is to use parallel processing. This can be accomplished either by free-space diffraction (FSD) gratings or by arrayed-waveguide gratings (AWGs). These grating technologies separate all channels in a single step rather than through a sequential series of filtering, which involves cost and complexity. There are several advantages to these technologies. First, these technologies permit high channel counts without requiring additional coupling devices, thereby resulting in a much smaller footprint, higher reliability, lower cost, and improved optical performance.

Also, although the arrayed waveguide grating technology is relatively new, it has demonstrated superior optical performance over the past few years, particularly with regard to insertion loss. Despite low insertion loss, these devices are outperformed by FSD grating technology. However, they suffer from higher PDL, thermal instability, and lower manufacturing yield.

FSD gratings also separate light in a parallel fashion and operate on the same physical principles as AWGs: induced phase shift and interference. FSD gratings are passive devices and do not require a heater to perform under all environmental conditions. Another major advantage of FSD gratings is their improved channel isolation that results from more accurate phase shift imparted by the grating. FSD grating technology offers better channel accuracy, lower PDL and lower chromatic and polarization mode dispersion. The last benefit is strictly due to absence of birefringence within the device.

Finally, FSD grating designs are often "single-ended," in which fibers enter and exit on the same side of the device. This permits efficient board layout, because it must accommodate only one fiber bend radius. Note all dispersive technologies produce a Gaussian passband response. When coupled with other devices having a Gaussian passband response, the effective filter function becomes narrow, which is not ideal for high-data-rate transmission. Flat-top versions of both technologies indicate that a desirable filter function is possible with a penalty of 2 to 3 dB in insertion loss. Such devices outperform the corresponding optical-performance parameters for high-channel-count subsystems using thin-film filters.

6.5.2 Low-Channel-Count Applications

Another DWDM application involves add/drop sites or nodes. Current network architectures employ fixed add/drop functionality, in which a particular node drops the same channels with no possibility of reconfiguration without physical intervention on site. In general, the number of channels dropped at a node varies from one to eight, but typical site drop varies from four to eight channels. Regardless of the channel count, the dropped wavelengths require demultiplexing from the network and a multiplexer to add channels on the main fiber. Thus, there is a significant need for low-channel-count (LCC) devices. Here the advantage of FSD gratings over thin-film filters and FBGs is less clear cut. LCC applications dilute the ability of FSD gratings and AWGs to amortize the fixed cost of the optics and packaging over a large number of optical fibers. Meanwhile, thin-film filters can be made with minimum cost and still can perform well in applications with fewer than 16 channels. Two emerging grating technologies have the potential to address this issue.

Dual-input FSD gratings developed by Zolo Technologies, Inc. share fixed costs of packaging and optics over two and potentially more multiplexing/demultiplexing devices housed in the same package. This reduces the cost per channel at least by a factor of two. This means a dual-input eight-channel device has approximately the same cost and performance level as a 16-channel device.

Another FSD grating technology that could improve the channel cost for low-channel application is called hybrid technology, which was developed by the National Research Council of Canada. The hybrid technology uses a curved, etched echelle grating that couples into and out of the device via planar waveguides. The hybrid technology has two potential advantages for LCC applications. First, the device has only two optical components, the grating and the waveguide coupler, thereby maintaining the device extremely cost-effective even for LCC applications. Second, the device is very small, which will significantly reduce the packaging dimensions. High yield on the etched gratings is of critical importance if minimum production cost is the principal requirement. Furthermore, the hybrid devices are not passive and require temperature control for proper operation, which will affect the cost and device reliability.

6.5.3 Functional Integration and Architecture

Future networks must use different but more flexible network architecture at the add/drop sites, with device-integration capability, if cost-effective design and minimum operating costs are the critical requirements. The network architecture must provide capability to bandwidth requirements according to the demand and must provide reconfigurable add/drop multiplexing and demultiplexing at the network nodes. Such technology requires a software provision to increase or decrease the number of channels to be added or dropped at any site and at any time. However, high-level, functionally integrated devices such as reconfigurable add/drops and dynamic channel equalizers will increase higher costs and complexity.

There are two problems with true monolithic integration using planar waveguide circuit technology. First, the yield for AWGs is very low. A reconfigurable add/drop multiplexer requires two 40-channel devices on the same substrate with a 2×2 optical switching technology displaced between the two waveguides for every channel. The yield for such a device is extremely low. Second, these devices would include a heater and components on a monolithic substrate with two low-yield devices. If one of the active devices fails, the entire device must be scrapped, leading to significant loss of investment.

Alternatively, functionally integrated FSD grating devices rely on a paradigm demonstrated by Zolo Technologies, Inc. This approach reuses the same grating and optics to perform two separate multiplex and demultiplex operations. Unlike AWG-based reconfigurable add/drops, FSD devices would require only a single grating, thereby realizing substantial savings. The switching technology could be a simple one-dimensional array of MEMS mirrors. The array requires as many mirrors as there are channels on the network. This type of array can be made relatively inexpensively and with good yields. The greatest benefit is that free-space integrated devices have the ability to switch out the failed active components during the manufacturing process, if not in the field.

6.5.4 Effects on Homowavelength Cross Talk

Consider a communications system comprising of drop fiber, ADD fiber (input fiber), output fiber, and four micromirrors in each of the four channels [7]. The tail portion of the beam that enters the optical add/drop multiplexer via the input fiber exits in the same fiber as the signal beam of the same channel that entered via the ADD fiber. Although the beam tail couples inefficiently into the fiber, it leads to a coherent, homowavelength cross talk. This type of cross talk presents a significant problem. An analogous situation occurs for tails that originate from either input fiber and that exit through either output fiber with main signal of the same channel originated from the other input channel. There are two performance issues of critical importance. The first is the amount of insertion loss as a function of beam-placement error. If this issue appears in the vicinity at a nominal wavelength of 1550 nm, the system performance is acceptable. In this region, there is favorable bias from a Gaussian mode approximation.

The second issue is the cross talk. The effect of cross talk appears in the region near the 1570 nm wavelength, which is the center wavelength of the neighboring channel. This wavelength value represents the portion of the power that leaks from the neighboring channel into the 1550 nm channel output. Even in a perfectly aligned system, a portion of the beam from each channel spills into the micromirrors of the neighboring channels. The exact amount of cross talk is a complicated function of finite beam size, nonideal system, and channel bandwidth. Furthermore, the amount of cross talk for a given system will vary in time because the state of the system varies with its environments and because the channel bandwidth varies with the rate of signal modulation. But it is clear that the Gaussian mode approximation produces significant amounts of error in modeling cross talk.

In addition, the amount of cross talk is affected by the relative states of the neighboring micromirrors. The main and tail portions of the beam for a given channel can be directed either to the same output fiber or to different fibers, depending on the position of the micromirrors. If the tail goes to a different fiber than the main portion, whichever portion of the tail that couples into the fiber will produce cross talk in the signals intended for that output fiber. If another signal of the same wavelength is in that output fiber, it will add to it as homowavelength or in-band cross talk. This type of cross talk presents a significant problem because it is from the same spectral channel as the beam from the ADD fiber and will therefore add coherently with the intended signal. Coherent summing or adding is based on amplitude rather than amplitude squared as in the incoherent case. If the phases from two optical sources vary, there can be large time-varying signal fluctuations even at very small level of cross talk.

6.5.5 Out-of-Band Cross Talk

If the tail ends up at a different output fiber than the main portion of the beam when no signal of the same wavelength is present in the output fiber, then the problem is not serious. This is referred to as heterowavelength or out-of-band cross talk. When the tail does not overlap with the other channels in the fiber, there is a good chance that it will be removed from the other signals during demultiplexing and will not show up as cross talk in the received signal. Even if it does show up in one of the other channel's final signal, it will sum up incoherently due to its different wavelength. Note signals of the same wavelength sum up coherently.

In summary, it is important to mention that system cross talk is one of the most important design issues for optical communications systems. It is imperative that the software model accurately determine its impact in the design of the system. Substituting a Gaussian beam mode approximation for the fundamental fiber mode can significantly underestimate the amount of cross talk. This, in turn, can lead to system components that will not meet the system performance requirements.

6.6 Optical Transceivers

Optical transceiver modules have potential applications in various systems including SONET, SDH, ATM, fiber distributed data interface (FDDI), FCFI, and GE [8]. These modules must meet the performance requirements for transmission, protocol standards, and optical connectors used in high-speed fiber-channel applications. Various local and international companies have designed and developed transceiver modules with exceptional system performance requirements while operating under severe environments.

6.6.1 Transceiver-Module Design Aspects

The transceiver module converts the incoming light waves to electrical signals and outgoing electrical signals back to light. The optical transceiver is based on solid-state laser diode technology. The transceiver module is a PCB, and the optical source is a tiny semiconductor chip: a light emitting diode (LED) or a laser diode. At optical frequencies in the near-IR region, the output of the solid-state laser can be modulated in tens of GHz leading to ultrawide bandwidth.

In an optical transceiver, a receiving port connects to incoming optical fibers, and a photodetector diode converts the light to electrical signals that are amplified, demultiplexed, and sent out for electrical interface. The photodetector requires an automatically power-controlled bias circuit (see Figure 6–9) to provide a constant operating voltage. Electrical clock and data-bit signals are synthesized, latched, and sent to the laser driver located in the transmitter module. The laser driver sends the signal as electrical current to the laser diode that converts the electrical energy to light energy. The laser driver must maintain a constant DC-bias current to set the laser operating point and must maintain the modulation current to carry the signal. To increase the signal throughput in a transceiver, the laser source must be carefully characterized for operating parameters to control the light-output intensity. In some transceiver designs, a feedback loop is used to stabilize the laser output power. However, the feedback loop complicates the transceiver design. Deployment of vertical cavity surface emitting laser (VCSEL) technology not only eliminates the feedback loop problem, but also does not require a photodetector.

6.6.2 Laser-Source Requirements

The laser is the most critical element of the transceiver module. A Fabry-Perot-type laser diode emits a coherent light beam from the narrow, beveled edge of the

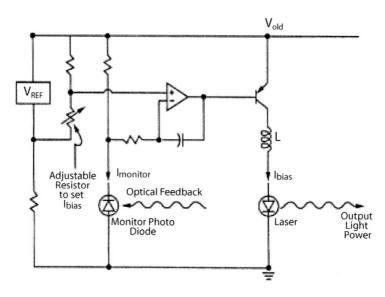

Figure 6–9 *Typical power control circuit using a monitor photodiode and adjustable resistor to set the bias control level.*

chip, with reflecting mirrors incorporated at the edges. However, a VCSEL laser is considered the most promising laser source. The VCSEL vertically emits the laser beam from a circular cavity with a diameter ranging from 5 to 25 micrometers, that is located at the top of the chip. The mirrors are incorporated as an integrated array on both ends of the cavity, and the design is known as a "distributed Bragg reflector." Deployment of parallel optical interconnects using multielement VCSEL arrays can provide a terabyte throughput capability.

Compared to edge-emitters, the VCSEL requires less current and has a lower lasing threshold current requirement (2 mA) compared to 30 mA for other solid-state laser diodes. The VCSEL's emitting aperture is very large, which means that the output beam divergence angle (a measure of dispersion) is significantly small. There are several manufacturing and processing advantages for VCSELs. The die is much smaller, which means that more VCSELs can be packed on a wafer with more interconnects, and all the VCSELs on the wafer can be tested at once. Furthermore, the VCSEL is more robust in operation compared to a laser diode, with a longer life expectancy and lower failure rates. In summary, a VCSEL design offers a low cost, lower threshold current, narrow output beam, lower failure rate, high reliability, and minimum power consumption.

6.6.3 Impact of Fiber and Detector on Receiver Sensitivity

SM-fiber transmission technology plays an important role in trunk transmission systems for future telecommunications systems. In the long-wavelength region, SM-fiber transmission systems provide both large transmission capacity and long repeater spacing because of low insertion loss and low dispersion characteristics of silica fibers. SM fibers with a relative index difference less than 2% and core diameter of less than 10 microns are best suited for fiber transmission systems. The material dispersion approaches zero near 1310 nm wavelength. A 1310 nm optical transmission system demonstrated a data-rate repeater product close to 90 Gbit/s-km as early as 1982. The 1310 nm dispersion-free fibers have a minimum insertion loss of 0.2 dB/km at 1550 nm wavelength, where the dispersion is close to 20 ps/nm-km. The 1550 nm fiber transmission system demonstrated a data-rate repeater-spacing product of 103 Gbit/s.km at 1550 nm wavelength. This indicates that the 1550 nm fiber transmission system offers the longest repeater spacing and largest transmission capacity in low-loss, silica fibers, provided dispersion-compensating devices are deployed to remove the transmission bandwidth degradation caused by the chromatic dispersion.

The sensitivity of the receiver is strictly dependent on the detector performance. An optical receiver using a p^+-n Ge avalanche photodiode (APD) as a photodetector has the worst sensitivity at 1550 nm wavelength compared to that at 1310 nm wavelength. Degradation in receiver sensitivity is caused by tailing in optical pulse response, which is dependent on the diffusion time for the carriers generated in the undepleted layer, and poor detector responsivity due to a reduction in the optical absorption coefficient. Receiver sensitivity at 1550 nm can be improved by using a reach-through-type p^+-n-n^- Ge APD detector. Note this type of detector has a longer depleted layer and thinner absorption length than a p^+-n Ge APD detector. A 1550 nm improved optical receiver using this type of detector and Si-bipolar transistor front offers data rates exceeding 2 Gbit/s with minimum cost and complexity. Note the detector with a longer depleted layer and thinner absorption length is known as the improved detector and is best suited for optical transceiver modules. The improved detectors are more resistant to temperature and bias voltage fluctuations. The quantum efficiency of the improved detector can be as high as 80%, depending on the bias level. The detector responsivity for a given data rate is dependent on the product of quantum efficiency and the pulse-response factor.

6.7 Optical Amplifiers for Communications Systems

Optical amplifiers are widely used by the LH transmission systems to compensate for the scattering and absorption losses in the optical fibers. EDFAs are best suited for LH-communications-system applications. The basic principle of the

EDFA states that the sum of signal photons, pump photons, and erbium-atom photons is equal to the applied signal photons [9]. When pumped by a 980-to-1600 nm laser, the energy level of erbium atoms jumps. Some atoms return to their ground state via spontaneous emission, producing a noise in the signal. However, most of them are knocked back to their ground state by signal photons with wavelengths somewhere between 1510 and 1600 nm. When this happens, the erbium ions emit a photon that has a wavelength identical to the signal, thereby providing a gain. EDFAs currently are used in communications systems to boost signals from 1530 to 1565 nm (C-band). Specialized optical fibers have extended the EDFAs to operate in the L-band region from 1570 to 1620 nm (see Figure 6–10).

In most communications systems, EDFAs with flat-gain response over wideband are preferred (see Figure 6–10). In certain applications, both the flat gain and low noise figure are required over the 40 to 60 nm spectral region. Note a WDM multiplexing scheme using an EDFA is more efficient and cost effective compared to a DWDM system because a DWDM system requires an EDFA with optical gain control and equalization features needed for a multiwavelength communication system. DWDM systems have been designed for 16, 32, and 40 channel capacities. DWDM systems with channel capacity of 100 or more are under development. However, future DWDM systems will require gain-locked, gain-flattened, optical amplifiers, regardless of the number of channels or signal allocation. In summary, low noise, high gain, low dispersion, and flat gain over wideband are the principal performance requirements for an EDFA, whether a communications system is WDM based or DWDM based.

6.7.1 Critical Performance Parameters of EDFA

EDFAs are deployed in three distinct locations in an LH transmission system. First comes the post-amplifiers located just after the transmitter end. The main function of a post-amplifier is just to provide momentum for the transmitted signals. Next comes a series of line amplifiers, located between each 80 km span of transmission fiber. These modules have stringent performance requirements for flatness, noise figure, and back reflections compared to post-amplifiers. Just before signals reach the receiver front end, the optical signals are given a final boost by a preamplifier to improve the SNR. If the signal level is too low compared to noise, then both the signal and noise are amplified by an amount equal to amplifier gain. Therefore, the EDFA offers flat gain over the 1530-to-1565 nm spectral region, which provides operation with minimum gain fluctuations from channel to channel. This characteristic of erbium could either extend the distance that permits single-channel transmission without regeneration or support additional channels. A single-stage EDFA offers limited gain, which may not be sufficient for a longer transmission distance.

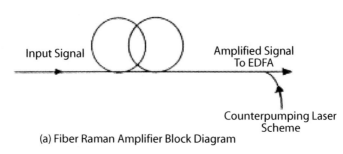

(a) Fiber Raman Amplifier Block Diagram

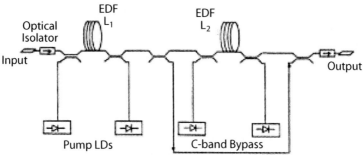

(b) Serial Wide-Band EDFA Topology Using Two Erbium-Doped Fibers (EDFs) and Multi-Pumping Technique

Figure 6–10 *(a) Fiber Raman amplifier block diagram and (b) schematic diagram of the serial wideband EDFA.*

Multistage amplifiers (see Figure 6–10) are best suited for a LH transmission system. In a typical configuration, a low-noise first stage amplifies the signals incorporating the 980 nm pump, strains them through FBG to smooth out the gain profile, and finally amplifies them in the second stage using a high-power 1480 nm pump. A typical 980 nm pump can pump as high as 300 mW, which is more than adequate for EFDA applications. The pump power requirement is slightly higher for a 1480 nm pump, whose typical output currently is close to 375 mW, according to leading telecommunications companies. Pump suppliers are predicting power levels for the 1480 nm pumps as high as 750 mW in the near future, which will yield higher amplifier output power to meet a large number of channels operating over a wide spectral region. The multistage approach permits the optical fiber to handle an increased number of channels in the C-band.

6.7.1.1 Impact of Gain Ripple

Gain ripple is the most important parameter of an EDFA. The amplifier gain curve shown in Figure 6-7 is less smooth outside the 1545 nm-to-1555 nm spectral region, where many new channels are operating on the line. Variations in per-channel gain are cumulative; therefore, the cascaded amplifiers incorporated in a transmission link will amplify not only the signals, but also the gain ripple between the wavelengths. This means that if an amplifier has maximum gain ripple of 1 dB, it will add up span after span. To avoid large gain ripple in a long transmission link, the individual amplifier gain ripple shall not exceed 0.25 dB between the maximum and minimum gain as per specifications. Gain-equalizing filters are used to reduce the intensity of the most amplified wavelengths so that they correspond with the lowest. In addition, the pumping source must provide more raw power to amplify the channel with the lowest gain.

6.7.2 EDFAs Operating in the L-Band

In the previous subsection, the requirements were summarized for the EFDAs operating in the C-band. Today's EDFAs typically support channels as many as 80. However, the demand for pump power from a single pump has significantly increased as the DWDM system has begun expanding into the 1570-to-1620 nm spectral range, as shown in Figure 6-7. This is called a long band or L-band, where EDFA gain begins to drop. The drop in erbium's gain occurs due to the fact that erbium has a low saturation threshold, which means that after a certain point, erbium will not respond to more pump photons. Furthermore, as the channel count approaches 100 and beyond, the amplifier's output can support so many dBs of each channel. Under this situation, the EDFA's saturation level and amplifier efficiency become critical operating issues. This leads to a conclusion that L-band amplification needs to explore other technologies to overcome the above problems.

The level of pumping affects the gain profile of the L-band EDFA. This means more power equals less bandwidth and vice versa. So it pays to increase the gain medium (or erbium atoms), not the pump power. This is possible by increasing the fiber length and/or by increasing the concentration of erbium dopant, as illustrated in Figure 6-11.

It is evident from the curves shown in Figure 6-11 that a concentration of 77 ppm in erbium dopant provides amplifier gain comparable to C-band gain. However, an increase in erbium concentration will result in higher attenuation, higher amplifier cost, and higher chromatic dispersion in the fiber. In addition, at high concentration, the erbium ions cluster together and emit as many photons as would an individual ion. Under these circumstances, the ions jump to higher levels and convert pump energy into signal. To overcome some of the above problems one can dope the fiber with relatively high levels of aluminum to distribute

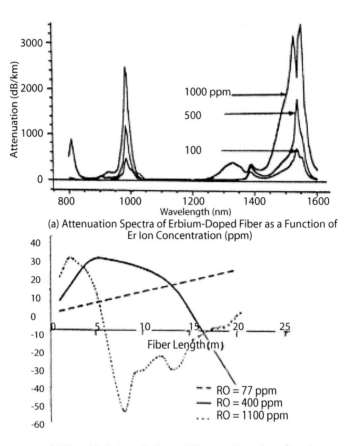

(a) Attenuation Spectra of Erbium-Doped Fiber as a Function of Er Ion Concentration (ppm)

(b) Signal Gain in an Er-Doped Fiber as a Function of Ion Concentration and Fiber Length

Figure 6–11 *(a) Attenuation spectra of Er-doped fiber and (b) signal gain as a function of ion contraction and fiber length.*

erbium atoms more efficiently through the fiber matrix. However, this approach will require fiber components with twice the length of those used in a C-band amplifier.

Currently, EDFAs are being used in LH transmission systems, serving as line, pre-, and post-amplifiers. However, expanding network capacity and lengthening transmission spans have imposed some constraints on per-channel power budgets and component costs. Network designers are looking for alternate technologies such as Raman amplifiers.

6.7.3 Raman Amplifiers

Except for silica's absorption peaks, Raman amplification is possible over an SM fiber's entire 300-to-2000 nm transmission spectrum. Amplification occurs when the pump photons couple with the vibrational modes of the fiber. The transfer of energy is released in the form of photons shifted roughly 100 nm towards the signal spectrum. In other words, a Raman pump emitting at 1450 nm will provide gain at about 1550 nm. Raman amplifiers are counterpropagating, which means that they amplify either approaching or departing signal photons. This amplifier architecture, shown in Figure 6.15, boosts approaching signals and directs the pump noise away from the receiver, thereby reducing its effect through fiber attenuation. Note a distributed Raman amplifier provides signal strength between 7 and 10 dB per fiber span, which translates into 35 to 50 km of extra distance, provided the loss in the transmission fiber does not exceed 0.2 dB/km.

A Raman amplifier assumes one of the two forms: a distributed Raman amplifier (DRA) or a discrete Raman amplifier (DISRA). The DRA module includes pumps, optics, and electronics and often functions as a line amplifier in conjunction with an EDFA. The DISRA module incorporates similar components as a DRA plus a span of fiber designed for a specific application. The DISRA is also known as a fiber Raman amplifier (FRA). The FRA is best suited for DWDM systems due to its distinctive flexibility in bandwidth design and growing maturity of high-power pump technologies. A single-wavelength-pumped FRA offers moderate gain. A multiwavelength counterpumped or backward-pumped FRA is best suited for a wideband Raman amplifier. Studies performed by the author on multiwavelength backward-pumped FRA (MWBP FRA) indicate that a 20 dB gain is possible over 1510 to 1610 nm with flatness not exceeding +/- 0.35 dB. The studies further indicate that minimum flatness is possible with a backward-pumping scheme deploying six pumps. Computed values of gain and noise figure for a backward-pumped FRA as a function of wavelength and number of pumps required are summarized in Table 6–2.

These computations assume a triangular Raman gain profile, identical pump losses, and zero photon energy loss. These assumptions will lead to a larger gain in the shorter wavelength band and a smaller gain in the longer wavelength region. Furthermore, these assumptions will yield larger errors in the shorter wavelength band than that in the longer wavelength band. These calculations are performed for the small signal case. For the large signal case, the signal gain is reduced approximately by 1.5 dB compared with the small signal case, due to gain saturation. However, the gain spectrum profiles obtained for both the small-signal case and larger-signal case are very similar in shape, which implies that they have nearly identical gain flatness.

Table 6-2 *Small-signal gain and noise figure (NF) for two MWBP FRAs.*

Wavelength (nm)	3-Pump FRA		6-Pump FRA	
	Gain (dB)	NF (dB)	Gain (dB)	NF (dB)
1510	16.5	6.2	19.5	7.5
1550	17.1	5.1	20.0	5.4
1600	19.8	4.0	20.1	4.1
1610	14.2	3.8	20.0	3.7

6.7.3.1 Gain Optimization Scheme

The trade-off studies performed by the author indicate that a gain-flattened optimization can be achieved using global optimization techniques of the genetic algorithm. This optimization scheme involves selection of a range for each pump frequency and input pump light power. The number of pumps selected must be very reasonable to meet cost-effective criteria. Optimum values of these parameters must be selected to provide uniform distribution of pump energy over the spectral range of interest. Optimum design of a MWBP FRA requires explicit knowledge of Raman gain coefficients between the various pumps involved, which can range from three to six to keep the cost down. Furthermore, the pumps must have identical triangle profiles to provide nearly identical Raman gain profiles. The fiber losses for all the pumps must be identical. Energy lost during the transfer from shorter-wavelength photons to longer-wavelength photons and signal-to-pump losses must be kept to a minimum.

6.7.3.2 Pump Parameters Optimization

Pump parameters optimization is of critical importance if gain flattening in a wideband optical amplifier is the principal design requirement. For FRAs, the gain flatness is dependent on the number of pump sources used, the pump power distribution of the optical sources, and the optical gain (G_n) of the nth signal in the band of interest. This gain is a function of pump light frequencies and the input pump light power levels. A closed-form analytical expression can be obtained for various pump power evolutions. In other words, application of a developed model for pump optimization design requires various pump power evolutions involving the Raman gain coefficient C_R between signal frequency (f_i) and pump frequency (f_j). For a triangle profile, the Raman gain coefficient is proportional to the difference between two pump frequencies ($C_R \propto f_i - f_j$). This coefficient is less than 13.3 THz [10]. Note pump power evolutions involve four basic assumptions: (1) The Raman gain coefficient is proportional to the difference between two pump frequencies and shall not exceed 13.3 GHz. (2) The optical fiber losses for pump lights are identical. (3) Energy loss whenever a short-wavelength photon is transformed into a long-wavelength photon must be neglected. This occurs when the ratio of two pump frequencies is close to unity. (4) The pump depletion caused by the signal-pump coupling must be negligible. Computations for optimum pump wavelengths and corresponding optimum pump power levels as a function of operating wavelength and number of pumps deployed in a Raman amplifier are summarized in Table 6–3.

An input power level of –40 dBm per channel for small signal gain and –2 dBm per channel for large signal gain are assumed for the following calculations in Table 6–3.

6.7.3.3 Performance Capabilities and Limitations of Raman Amplifiers

Raman amplifiers require sufficient output power levels from solid-state laser sources to stimulate the Raman effect in the transmission fiber. Early Raman amplifiers to amplify optical signals employed double-clad ytterbium-doped fiber pumps capable of delivering output power close to 3 W. Applications of these pumps continue today in short underwater optical links to avoid using expensive undersea-qualified erbium-doped amplifiers. Within the last two years, solid-state diode-laser pump power levels have caught up sufficiently to allow Raman amplifiers to leave behind the Yb-doped fiber and elicit amplification on their own.

Table 6–3 *Optimum pump wavelengths and corresponding pump power levels for three-pump and six-pump schemes.*

3 –Pump Scheme		6-Pump Scheme	
Wavelength (nm)	Pump Power (mw)	Wavelength (nm)	Pump Power (mw)
1423	1350	1404	680
1454	190	1413	600
1484	200	1432	440
----	----	1449	190
----	----	1463	76
----	----	1495	54

A typical distributed Raman amplifier (see Figure 6–10) consists of two pairs of SM, 250 mW diode lasers, each pair designated to a specific wavelength, with pairs combined through a polarization multiplexer. The twin-pump design provides more pump power. A Raman amplifier using this pump scheme delivers output power even with coupling losses in excess of 500 mW and a gain ranging from 10 to 12 dB. In addition, separating the pump wavelengths by 25 to 40 nm helps to achieve a flat gain curve in a single-band amplifier's 32 nm spectral bandwidth. Note the spacing between the pumps is dependent on the type of optical fiber used, but typically one can use 1427 nm and 1467 nm pumps to provide adequate gain in the C-band amplifiers. It is important to point out that major cost for Raman amplifiers is from high-power, solid-state diode-laser pumps. However, as far as high amplifier output is concerned, other optical amplifiers cannot outperform Raman amplifiers.

6.7.4 Performance Limitations of Amplifiers in Optical Networks

The EDFA has revolutionized the optical signal amplification technology and becomes a critical component to amplify WDM signals in optical networks. In the case of DWDM technology, different channels are multiplexed in the wavelength domain with interchannel spacing of 0.8 nm. When these different channels propagate through the EDFA for signal amplification, they suffer from interchannel cross talk resulting in significant reduction in the SNR due to uneven gain spectra of the EDFA [9].

Note the EDFA offers a fiber cost-effective option for periodic amplification of optical signals in the low-loss regime of the silica-based fiber. For a single-carrier transmission, the EDFA offers the most cost-effective approach to operate transmission distances exceeding a few thousand kilometers without use of a repeater. However, for WDM systems, the use of an EDFA presents some system-orientated problems.

The EDFA gain is dependent on the pump power and coupling losses. The amplifier gain decreases linearly as the input-signal power is increased. The EFDA gain increases with the increase in pump power. However, minimum gain variations occur at lower pump power levels, regardless of the operating wavelength. Amplifier signal gain and output power as a function of fiber length and forward pump power (FPP) are illustrated in Figure 6–12.

6.7.5 Multichannel Amplification

Requirements for multichannel amplification are quite different from those for single-channel operation. Consider two channels operating different wavelengths and transmission data rates. Channel I uses a wavelength of 1550.2 nm for a data rate of 155 Mbps, and channel 2 uses a wavelength of 1551.9 nm for a data rate of 155 Mbps. Figure 6–13a shows the output-signal power level as a function of pump power for Channel I with and without Channel II. The drop in output power when propagated with Channel II indicates that the interchannel cross talk affects the SNR at the output of the EDFA (see Figure 6–13b).

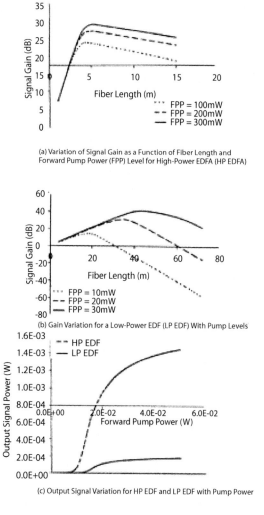

(a) Variation of Signal Gain as a Function of Fiber Length and Forward Pump Power (FPP) Level for High-Power EDFA (HP EDFA)

(b) Gain Variation for a Low-Power EDF (LP EDF) With Pump Levels

(c) Output Signal Variation for HP EDF and LP EDF with Pump Power

Figure 6–12 *Amplifier gain variation as a function of pump power levels for (a) HP BDFA and (b) LP EDFA fibers and (c) output power.*

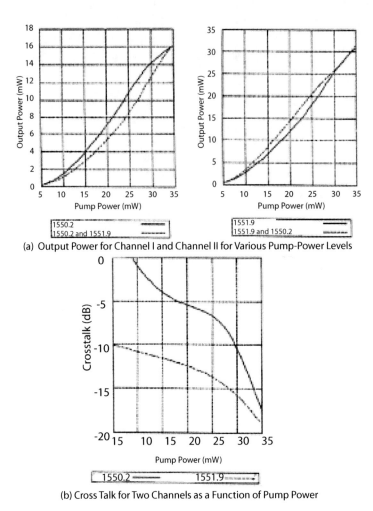

(a) Output Power for Channel I and Channel II for Various Pump-Power Levels

(b) Cross Talk for Two Channels as a Function of Pump Power

Figure 6–13 *(a) Output power for channels and (b) cross talk for two channels as a function of pump power level.*

When two channels propagate together through the EDFA, there is CT between the channels. The CT introduced in the channel can be defined as

$$CT = [P_i - P_c]/P_i \qquad 6.5$$

where P_i is the power in the channel when it propagates in a single-channel mode along the pump, and P_c is the power in the channel when copropagated with other

WDM channels, as illustrated in Figure 6–13c. Note as the pump power is increased the CT in both the channels decreases and is around –18 dB for a pump power of 35 mW. The CT is more than –5 dB when the pump power is reduced below 20 mW. This means the higher the pump power, the lower the CT will be.

The effect of adding/dropping channels is illustrated in Figure 6–13b. The output signal power and CT in the 1550 nm channel are plotted as a function of number of channels added. This shows that the decrease in output signal power and increase in CT are directly proportional to the number of channels added to the one-channel system.

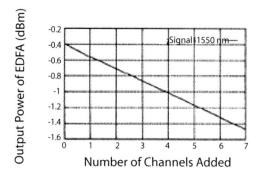

(A) Output Power of Channel I as a Function of Channels Added

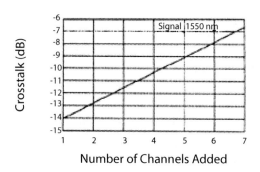

(B) CT in Channel I as a Function of Channels Added

Figure 6–14 *Power output and CT in Channel I as a function of number of channels added to the WDM system.*

6.7.6 High-Power EDFA Requirements

Studies performed on high-power EDFAs reveal that Er^{3+}-doped silica fiber with a higher numerical aperture (NA) and higher erbium-ion concentrations are the principal requirements in the design of a high-power EDFA (HP EDFA) operating in the 1550 nm spectral region. The studies further reveal that optimum fiber length in an HP EDFA does not change appreciably with the increase in pump-power or input-signal power levels. HP EDFAs are best suited for applications in the fields of nonlinear optics and multigigabit/s WDM signal generation. In multigigabit/s signal generation, the signal power from the EDFA must be in excess of 300 mW for demultiplexing into several channels after spectral enrichment. A 980 nm pump is generally selected because it offers higher gain, which is measure of quantum conversion efficiency (QCE) of the pump. The QCE is dependent on output-signal level, input-signal level, and input pump photon flux levels. The QCE can be improved by 60% just by increasing the fiber NA from 0.15 to 0.25. A further increase by 20% in QCE is possible by increasing the erbium ion concentration. Fibers with an NA of 0.22 and ion concentration of 3.168×10^{24} ions/m^3 (or 400 ppm by weight) are best suited for HP EDFA applications.

In the case of an HP EFDA, gain in excess of 30 dB is possible with pump power of about 30 dBm, even for fiber lengths as short as 5 meters. This is due to an increase in QCE. In contrast to a low power EDFA (LP EDFA), the output-signal power does not saturate, even for 50 mW of pump power. The optimum fiber length is a function of erbium-ion concentration because both the pump absorption and signal gain coefficients are proportional to it. The optimum fiber length decreases to 5 meters as the ion concentration is increased to 3.168×10^{24} ions/m^3 (or 400 ppm by weight). The optimum fiber length is larger than 20 meters as the ion concentration is reduced to 6.1×10^{23} ions/m^3 (or 77 ppm by weight). Note at ion-concentration levels of 900 ppm and higher, the pump depletion is so fast that after a short length of erbium doped fiber (EDF), one will notice the reverse transfer of power from signal to pump, as illustrated in Figure 6–11. Therefore, it is very critical to select the length of the fiber appropriate to the value of ion concentration in the fiber.

In summary, an increase in NA and ion concentration leads to an increase in QCE, thereby leading to an increase in output-signal power and decrease in the required optical length of the HP EDF. Furthermore, the optimum length of the HP EDF does not vary significantly with the change in pump power or signal power. This is in contrast with an LP EDF, where the optimum length of EDF increases almost linearly with pump power.

6.7.7 Optimum Topology of Wideband EDFA for WDM Applications

EDFAs operating in the long-wavelength region (L-band) over the 1570-to-1610 nm spectral region have doubled the capacity of DWDM communications systems. The principal objective of reconfiguring the EDFA for optimum performance is to achieve another 40 nm gain window. Over the last three years, gain-flattened L-band EDFAs have been widely used for WDM systems to achieve cost-effective operation [10]. The dramatic increase in transmission capacity of WDM systems has been achieved by incorporating a parallel configuration of gain-flattened conventional C-band and L-band EDFAs. Incorporating attenuators in front of the WDM multiplexer to equalize the flat gains of both these EDFAs has provided a new broadband EDFA topology that yields an impressive gain of 18 dB over a 75 nm band with total input power of –5 dBm. This serial topology comprising of two stages combined is shown in Figure 6–10. Two potential topologies, the serial topology and parallel topology, are available and can each be optimized for codirectional and bidirectional configurations. Finally, the parameters of the serial technology for WDM-system applications can be optimized and can be compared with the parallel configuration of gain-flattened C-band and L-band EDFAs.

An appropriate numerical model can be used to optimize the amplifier performance involving pump power output, multiple input signals, and spectral components of forward and backward propagating amplified spontaneous emission (ADE) power levels along the EDF. Atomic population densities at the ground level and detestable energy levels can be calculated by solving the rate equations. The numerical model computes the spectral power density of ASE, which leads to the determination of the noise figure of the amplifier. The numerical model can optimize the parameters of the first and second stages: EDF lengths L_1 and L_2, pump configurations (see Figure 6–10), and pump power levels to achieve a minimum gain of 20 dB for as many C-band and L-band channels as possible. However, this kind of gain performance requires minimum values of intrinsic gain ripple and noise figure.

The gain ripple of WDM signals is defined in terms of minimum and maximum gain and can be written as

$$G_R = [G_{max} - G_{min}]/G_{max} \qquad 6.6$$

where G_{max} is the maximum gain and G_{min} is the minimum gain.

Preliminary calculations indicate that a gain of 20 dB/channel can be achieved with acceptable ripple for the 32 C-band and 40 L-band channels. The lowest noise is only possible when both the first and second stages are pumped codirectionally at 980 nm wavelength and with pump power levels of 145 and

180 mW, respectively. The calculations further indicate that the optimum EDF lengths are 10.5 and 80 m for the first and second stages, respectively. Note the gain ripple and interchannel power spread in a multiwavelength regime are very sensitive to the position of the first WDM signal. The calculated values of gain ripple at the output port of the serial topology are 0.048 and 0.047 over the 32 C-band and 40 L-band channels, respectively. Identical gain ripple can be achieved for other pump configurations. Note the addition of any amount of counterdirectional pump power at 1480 nm will result in the highest increase in noise figure for both the C-band and L-band channels. In summary, the new technology for serial configuration can provide gain of 20 dB for the 32 C-band and 40 L-band channels with 0.8 nm spacing and input power of −17 dBM/channel with gain ripple less than 0.05 or 5%. The lowest noise figure occurs when both the first and second stages are pumped codirectionally at 980 nm. In comparison with the usual parallel configuration of C-band and L-band EDFAs, the first EDFA using new technology requires about 20% more pump power compared with the C-band EDFA of the parallel topology because it provides gain only from 7 to 15 dB for the L-band signals. The optimum EDF length and pump power of the second stage are smaller by 10% and 33%, respectively, compared with the L-band EDFA of the parallel topology. In brief, the new topology saves about 20% of the overall pump power and 10% of EDF length and achieves a lower noise figure in L-band channels than the parallel configuration, thereby realizing significant reduction in the WDM systems cost.

6.8 Summary

Performance requirements for the FO-based components and devices widely used in WDM, LH DWDM, and metropolitan WDM communications systems are identified. Advantages of FO communication links for optical control of transmit/receive modules in phased array tracking radars are summarized. Beamforming techniques for microwave and optical domains are discussed. The impact of nonlinear effects on high data rates in LH transmission systems is identified. Performance requirements for critical elements of EDFAs and Raman amplifiers are summarized. Important component parameters affecting the out-of-band and in-band CT are identified. Potential techniques to boost the operating bandwidth of optical amplifiers are described. Pump power requirements for EDFAs and FRAs are identified with emphasis on gain-flattening and bandwidth enhancement. Advantages and disadvantages of serial and parallel configurations for WDM applications are summarized. Various approaches to achieve optimum gain and minimum gain ripple over wide spectral regions in EDFAs are identified. Techniques to achieve optimum small-signal gain and a minimum noise figure in multiwavelength, backward-pumped FRAs are summarized as a function of wavelength, pump power level, and pump configuration.

6.9 References

1. Jha, A. R. (2000). *Infrared technology: Applications to electro-optics, photonic devices, and sensors* (pp. 311–313). New York: John Wiley and Sons, Inc.
2. Senior Editor. (1998). Optical control of microwave components and devices. *Microwave Journal*, 310–320.
3. Dunay, N. G. (2000, November). DWDM permeates networks from access to ultra long-hauls. *WDM Solutions*, 19–20.
4. Hecht, J. (2000, October). Long-haul DWDM systems go the distance. *Laser Focus World*, 125–132.
5. Survaiya, S. P., et al. (1999). Dispersion characteristics of an optical fiber using linear chirp refractive index profiles. *IEEE Journal of Light Wave Technology*, *17*(10), 1797–1805.
6. Sappy, A. (2002, May). Not all multiplexing technologies are on the same wavelength. *Photonics Spectra*, 78–84.
7. Shiefman, J. (2002, May). cross talk in multiwavelength optical cross-connect networks. *Photonics Spectra*, 130–133.
8. Fedele, P. (2001, March). Behind the light show in optical transceivers. *ISD Magazine*, 28–31.
9. MacCarthy, D. C. (2001, July). Report on fiber amplifiers. *Photonics Spectra*, 88–93.
10. Narayankhedhar, S. K., et al. (2000, December 18–20). *Performance evaluation of WDM optical networks with erbium-doped fiber amplifiers.* Presented at PHOTONICS-2000 International Conference, Calcutta, India, pp.261–264.
11. Karasek, M., et al. (2001). Serial topology of wideband erbium-doped fiber amplifiers for WDM applications. *IEEE Photonics Technology Letters*, *13*(9), 939–941.

CHAPTER 7

Fiber Optics for Medical and Scientific Applications

This chapter is dedicated to the applications of FO technology in medical and scientific fields. In the past two decades, FO technology has played a key role in medical and scientific research. FO-based lasers are finding potential applications in medical diagnosis, dental surgery, and other bloodless surgical procedures. The laser is the critical element of flow cytometry, which permits a large light flux to be coupled into a small area through low-loss optical fibers capable of producing sufficient fluorescence intensity for instant and accurate analysis of a single cell. FO-based catheters are widely used for medical diagnosis involving angioplasty, colon testing, and angiography. Q-switch laser pulses are delivered through a novel FO catheter to emulsify the blood clot associated with an ischemic stroke. This laser treatment, at a wavelength of 532 nm, can play an important role in eliminating the blood clot in a cerebral artery, thereby providing a new lease on the patient's life.

FO technology is widely used by various lasers, including gas lasers, solid-state lasers, and ultrafast lasers to conduct biomedical studies and clinical research. Special types of optical fibers are used to deliver the laser energy for specific medical applications. High-quality quartz optical fibers are best suited for curing various dental problems. A FO-based microarray can detect and test DNA samples, involving 50,000 individual optical sensors. During the last two decades, FO lines have been widely employed by mini-IR cameras to detect ulcers and other gastronomical diseases. Special lasers and low-loss, low-dispersion optical fibers are best suited to perform a variety of surgical procedures in the fields of ophthalmology, general surgery, dermatology, dentistry, laser angioplasty, and veterinary medicine.

High-power solid-state lasers using AlGaAs (810 nm), InGaAs (960 nm), Nd: YAG (1060 nm), and diode bars have potential applications in medical fields, including urology, gynecology, ENT-treatment, ophthalmology, photodynamic therapy (PDT), laser disc compression, liver tumor ablation, and angioplasty. Excimer lasers also play a key role in specific medical treatments. A mix of argon and krypton gases in correct proportion is required in the design of an excimer laser. A typical CW power levels of 1 W is available for excimer lasers operating in the 350-to-530 nm spectral region, but peak energy levels close to 1μJ are possible when operating in the pulse modes. Procurement costs vary from $8,000 to $12,000 for a CW excimer laser, and from $16,000 to $20,000 for a pulse excimer laser depending on the emitting wavelength and the power-output level.

7.1 Potential Techniques for DNA Analysis

Several IR and electro-optic techniques are available for DNA analysis, including ion mobility spectrometry, capillary electrophoresic processing, and fluorescent fine-narrowing spectroscopy. However, the detection of the signal is accomplished by a sensitive detector array or high-resolution camera based on a charged coupled device (CCD). Regardless of the technique employed, DNA analysis requires comprehensive knowledge and state-of-the art clinical instrumentation.

7.1.1 Ion Mobility Spectrometer Technique

Ion mobility spectrometers can be successfully used to trace residues of tear gas. These sensors are best suited for riot-control environments and for maintaining civil order during riots or unlawful disturbances. Low-tech devices such as polarized light microscopes (PLMs) can be used to identify hair and clothing fibers of the rioters, thereby pinpointing the criminals who took part in the riots. In certain cases, these devices can identify explosives, but the operator must possess comprehensive knowledge and extensive experience to recognize specific crystal structures. Identification of hairs or fibers is a leading technique to link a potential suspect to a crime and the results obtained using this technique may be admissible in the court. A more obvious link is possible when a perpetrator leaves behind a tiny sample of blood or semen. Forensic scientists or biologists can generate a DNA "fingerprint" of the perpetrator even from the tiniest sample collected at the crime scene [1]. Forensic scientists consider these DNA tests more reliable, and they can either rule out a suspect or strengthen the prosecutor's case.

7.1.2 Laser-Based DNA (LBDNA) Analysis

The LBDNA technique, also known as the capillary electrophoresis (CE) process, is currently receiving a lot of attention. In this process, forensic scientists extract the DNA samples and create a sort of carbon copy of each strand with a specific number of tandem repeats, which are then treated with fluorescent tags to highlight their unique number. This number is necessary for conclusive and meaningful DNA analysis. An argon laser emitting at 488 nm or 514 nm is used to shine directly on the tiny glass capillaries, causing the DNA to fluoresce. A CCD array with 640×512 pixel resolution capability automatically records each sample with high accuracy. This laser-based system allows a laboratory technician to perform a reliable DNA analysis on smaller samples of generic material. In addition, the LBDNA equipment is capable of analyzing samples that have been degraded or are few in number. An LBDNA system using a laser source with 20 mW at a wavelength of 532 nm can record the fluorescence of a DNA band with two photomultiplier tubes that have much higher gain than conventional avalanche diode photodetector (ADP) devices. The technicians can switch optical filters and fluorophores to mark definite characteristics on the DNA strands with high resolution. The CE technique requires a high degree of skill in operating the system and extensive experience in proper loading of each sample. Representing the latest photonic technology, this LBDNA technique is taking a bite out of crime. It is important to mention that a high-energy laser source and optical fibers with low insertion loss are required to achieve 3-D electropherograms for the blood samples collected from the scene.

7.1.3 Fluorescent Line-Narrowing Spectrocopy (FLNS) Technology for DNA Use

DNA analysis offers a powerful tool for law enforcement agencies and district attorneys to prosecute criminals. DNA samples and other-fluorescent analytics generally have been embedded into the glass, which can analyze the low-temperature, FLNS technology for precise DNA analysis data. Selective excitation at the liquid helium temperature of 4.2 K provides high-resolution spectra [1] of the samples that can be used in fingerprint identification. The CE technique with FLNS technology for on-line spectral characterization represents a source of vital information for rapid detection that is capable of providing reliable structural information on the samples collected. A CE-based FLNS system offers an additional degree of selectivity in the chemical analysis of chemical compounds with similar structures. Employing the nonselective, high-energy laser excitation and automatic translation of the system through laser excitation at 77 K temperature, 3-D electropherograms can be obtained with relevant characteristic fluorescence spectra that is capable of determining the fluorescence original bands known as (0,0) bands. Once the original band transitions are known, appropriate laser excitation wavelengths can be selected for the low-temperature (4.2 K) FLNS characterization of the DNA sample analysis.

7.1.4 Laser Excitation Requirements for DNA Analysis

DNA specialists and clinical scientists have found an excimer-laser pumped dye the most suitable for laser excitation. FLN spectra are generated using a series of laser wavelengths that selectively excite [2] vibronic regions of the first excited singlet state of the compound, each of which reveals a portion of the excited-state vibrational frequencies of the molecule. Fluorescence is then collected at a right angle to the laser excitation beam, dispersed by the monochromator, and detected by a sensitive photodiode array or high-resolution CCD-based camera. High-resolution FLN spectra obtained by the FLNS technique using different laser excitation wavelengths can be used as "fingerprints" for spectral identification of a complex compound such as blood or a single element such as hair. This DNA analysis has potential application in biological research and forensic science.

7.1.5 Optical Fiber Requirements for DNA Analysis

Stringent requirements have been imposed on optical fibers for use in DNA analysis to achieve reliable and authentic data. Detecting very small quantities of chemical or biological samples or material for security and defense applications is fraught with serious problems, including the generation of false positives and negatives as a result of insensitivity to or accidental contamination of the samples. Studies performed by scientists at Tufts University (Boston) indicate that deployment of a FO microarray consisting of 50,000 optical fibers will permit detection of samples by 50,000 individual sensors simultaneously. The studies further indicate that an optical fiber can be melted and pulled into a capillary with nanometer (10^{-9} m) dimensions. This will provide resolution and sensitivity of the highest order for DNA analysts. Thousands of optical fibers are melted and pulled simultaneously to produce a thick bundle of individual light pathways, each connected to a separate sensor. A bundle configuration shown in Figure 7–1 has an area of 1 mm^2 that can contain as many as 50,000 optical fibers, each having a diameter of 3 microns or 0.00003 mm. This permits an optical fiber microarray with the highest-possible packing density.

Using a hydrogen-fluoride (HF) solution, the end of each fiber can be exposed to make an etched well approximately 3 microns deep that can accommodate sensors such as DNA strands, bacteria, or yeast cells. All these sensors are embedded with dyes capable of yielding an optical code bar to identify the position of each sensor. When a sensor contacts a particular biological agent or its labeled DNA, the optical microarray system measures a fluorescent signal. Because each sensor is coded with dye, the DNA analysts will know which sensor has been activated. DNA scientists claim that incorporating three dyes at five concentration levels provides the potential for hundreds of combinations, leading to reliable DNA analysis.

An arc lamp, a dichroic housing, and a high-resolution CCD camera are required to read the sensor, using the microarray with closely packed arrangement of 50,000 optical fibers, each having a diameter of 3 microns. In the opinion of the research scientists, the detection limits are extremely low. A few molecules in one of the 3-micron wells produce a very high local concentration. However, this problem can be solved by using the latest version of this approach, which involves a cell-based sensor capable of detecting changes in numerous types of bioagent spores or DNA-damaging agents.

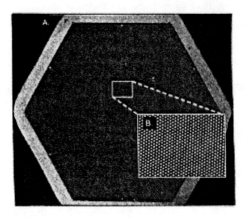

A: Represents a bundle of 50,000 high-performance optical fibers capable of providing remarkable sensing capability enclosed in a compact space.

B: Represents a magnification of the hexagonal bundle showing the closely packed arrangement of 50,000 optical fibers, each having a diameter of 3 microns.

Figure 7–1 *Optimum configuration of FO microarray for DNA analysis.*

7.2 Laser and FO Technologies for Dental Treatment

Solid-state lasers and low-loss FO delivery systems are widely used in treating various dental diseases. Laser-based spectrometers are playing an important role in dental surgical and diagnostic procedures. Studies performed by the author indicate that erbium-doped YAG (Er:YAG) lasers emitting at 2.94 microns and semiconductor lasers are best suited for diagnosis of hard-to-detect dental diseases. The studies further indicate that Nd:YAG laser-based Fourier transform Raman spectrometry (FTRS) is most effective in the detection of tooth decay. According to dentists, the areas of decay with significant bacterial activity are highly fluorescent. This means that Raman scattering, using Nd:YAG laser-based FTRS, could prove effective for locating hard-to-detect infection beneath the tooth surface. Although advanced tooth decay is easy to distinguish on healthy tooth enamel, it is much more difficult to detect in infected dentin, beneath the tooth enamel [3]. Clinical scientists believe that the ability of Raman spectroscopy to locate the underlying infections and distinguish them from the healthy tissues could result in a powerful diagnostic tool for the dentists and dental surgeons. A Raman spectroscopic microscope, illustrated in Figure 7–2, is effective in treating various dental diseases. This microscope consists of an argon laser pumped with a Ti:sapphire laser. Some clinical scientists think that a near-IR

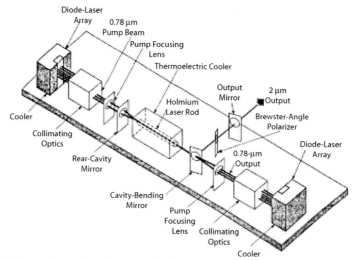
(a) Holimum Laser Configuration for Treating Heart Ailments

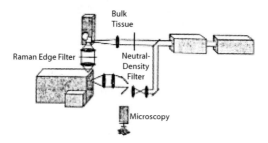
(b) Raman Spectroscopic Microscope/Tunable Ti:sapphire Laser

Figure 7–2 *Holium-based laser and Raman spectroscopic microscope for transmyocardial revascularization (TMR) procedures. Courtesy of John Wiley and Sons, Inc., New York.*

tunable laser operating over the 750-to-850 nm spectral range will be effective in collecting Raman spectral signals more rapidly. This particular tunable laser has reduced the signal collection time from 30 minutes to less than 5 seconds.

7.2.1 Laser Requirements for Dental Applications

Laser wavelengths with water-absorption peaks must be evaluated for dental applications. Dentists feel that lasers operating at 3 microns are best suited for dental procedures involving bone cutting and tooth drilling. Erbium-doped lasers such as Eryngo and Er:YSGG are widely used in medical diagnostic applications. The 2.94-micron Er:YAG laser is closer to the nominal water-absorption peak than the 2.79-micron Er.YSGG laser absorption line. Flash-lamp pumping schemes can be used for these lasers. These lasers are codoped with chromium to improve pump light absorption performance. Flash-lamp pumped Er.YAG lasers generate CW power as high as 30 W, whereas a singly-doped erbium diode-pump laser at 970 nm provides output power levels exceeding 500 mW. This is more than sufficient for hard-tissue applications [4].

Laser techniques to eliminate dental fillings are being investigated. A new laser technique developed by the University of Glasgow [5] is based on two-photon absorption from a solid-state Cr:LiSAF laser, which monitors the changes in laser-excited fluorescence that occurs as decay attacks the enamel coating of the healthy teeth. Dental tissue contains natural fluorescence that absorbs the laser energy between 420 and 450 nm wavelengths. In a healthy tissue, the peak of the fluorescence occurs at 680 nm, but in diseased enamel, it occurs at 550 nm. This indicates a significant difference between healthy and diseased enamel. The two-photon laser imaging technique provides the depth discrimination with minimum light loss because of the significant resolution provided by the two-photon absorption phenomenon.

The Cr:LiSAF laser provides early detection of tooth decay, thereby eliminating unnecessary fillings or dental drilling to remove diseased tissue. This laser provides 100-femtosecond (10^{-15} second) pulses at the repetition rate of 200 Hz to maintain both the average and peak power levels at the optimum values. However, the laser light focused on the tooth under investigation is in the order of a few milliwatts. A red aiming laser beam can be used to achieve sharp images of the tooth, locating the "fluorescent hot spots" on the tooth surface. Essentially, this particular laser is best suited for prevention of tooth-related problems.

7.2.2 Specialty Fibers for Medical and Dental Applications

Conventional optical fibers are not suitable as laser delivery systems operating around 3 microns or so. Optical fibers with special performance requirements are used in medical diagnostic and dental surgical procedures. Special fibers are available in a wide range of core diameters best suited for a specific range of operating wavelengths. These optical fibers come with various numerical apertures compatible with mode-field diameter and bend-sensitivity requirements. Characteristics of optical fibers best suited for dental applications and for specific spectral ranges are summarized in Table 7–1.

The optical fibers shown in Table 7–1 provide optimum performance in the spectral regions specified. The material selected for the optical fiber yields excellent durability and high-quality performance under variable ambient environments. Precision armor jacketing and autoclavability are necessary for these fibers to retain high optical performance. High-quality quartz cores are used for UV curing in various dental applications.

Table 7-1 *Characteristics of optical fibers for dental applications.*

Fiber Core Type	Wavelength Range (nm)
Plastic	400–900
Boro Silica	400–1000
Quartz: (UV to Visual)	193–1200
Quartz: (UV to IR)	350–2400
Fluoride	2000–4500

7.2.3 Optical Illuminators

Optical illuminators are required in certain dental applications where intense cool light is critical. Optical illuminators can be integrated with the optical fiber delivery systems to provide cool, clean, and focused light when and if required in certain dental surgical procedures for the comfort of the patient.

FO light sources integrated with high-efficiency reflector lamps are suitable for use as built-in light sources in FO-based systems applications. These sources provide light intensity ranging from 0.5 to 150 W. Dichroic reflectors designed for high throughput allow emission wavelengths greater than 700 nm to pass through with minimum heat, thereby selecting only the visible light transmission. Ellipsoidal aluminum reflectors with diameters ranging from 9.5 to 50 mm produce a small, uniform, high-intensity spot for full-spectrum efficiency. The lamps can be bonded permanently into the reflector to create a prefixed spot if needed for special dental procedures.

Randomized light guides and ring lights, as shown in Figure 7-3, are best suited for special dental applications. Randomized light guides provide even light distribution across the face of the fiber area, as illustrated in Figure 7-3. This light guise makes for an even transmission of any type of light such as incandescent, IR, UV, etc. Single and bifurcated light guides are available in any configurations. End tips, lensing, and jacketing can be provided to meet required performance specifications.

Ring lights, as shown in Figure 7-3, are best suited to provide a full 360° annulus of intense, white light. They can be designed for use with stereomicroscopes. High-transmission FO ring lights are commercially available in any length. Ring lights can be designed with a crush-resistant stainless-steel interlock or PVC monocoil jacket capable of operating under harsh operating environments.

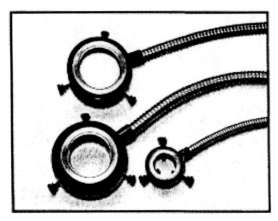

(a) FO-Based Optical Illuminators Capable of Providing Cool and Focused Light

(b) Randomized Light Guides to Provide Even Light Distribution

Figure 7-3 *(a) Optical illuminators (b) and randomized guides.*

7.3 Spectroscopic Technology for Life-Science Research

Hyperspectral-imaging spectroscopic (HIS) technology offers a unique imaging technique for identifying and quantifying the relationship between biologically active elements, enabling the acquisition of spectral information with high reliability. Traditional spectrometers [6] acquire the spectra of homogeneous materials. If the sample is a liquid mixer, then the scientist uses chromatography or any other appropriate separation technique to purify the components and then analyze the results. If the sample is a tissue or cell specimen, it is possible to separate the constituents without changing their molecular structure, which is required for a thorough and reliable analysis.

In the case of the HIS technique, the HIS instrument acquires the individual spectrum of each cell component and assumes that the material under investigation is heterogeneous. The HIS instrument then identifies spatially resolved objects based on their spectral signatures. This technique requires up to 200 spectral data points per spectral object, whereas the multispectral imaging (MSI) technique requires a maximum of 20 data points per spectral object. Using the HIS technique, a research scientist can simultaneously obtain spectra of complex multiple overlapping fluorophores and separate them through deconvolution algorithms. In both the HIS and MSI techniques, scientists need to obtain the spectral and spatial information with high reliability. Spatially resolved spectrum-acquisition devices can be designed based on two operating concepts: the spectral-cube and image-cube concepts, as illustrated in Figure 7–4.

An image cube acquires the entire spectrum from 365 to 750 nm simultaneously and generates images sequentially with one slice at a time across the field of view (FOV). In this method, the sample scans across the entrance slit of the spectrometer. A spectral cube acquires a fixed FOV layered in sequentially acquired wavelength slices. Each slice is identified by its own specified wavelength ranging from λ_1 to λ_n over the 365-to-750 nm spectral range, as shown in Figure 7–4.

This means that the HIS instrument builds one spectrum at a time. Each pixel in the IR CCD-based camera produces a complete spectrum defined by the number of wavelengths acquired. The cubes resulting from the two distinct methods present spatial and spectral data in two different ways. Spectral cubes can be created by various tunable filters, including acousto-optic filters, dielectric filters, and liquid-crystal filters. The spectral cube method is best suited for locating fluorophores dispersed in the cells or tissues. The spectral cube method takes a long time. If the spectra within the sample changes under various environmental conditions such as temperature, pressure, or humidity, the information from the same sample may be different at the first and last wavelengths. Furthermore, the huge size of a spectral-cube file will slow the processing speed and thus delay the receipt of data. One can estimate the file size per wavelength and the overall file size for the number of wavelengths involved. Assuming a spectral coverage of 400 nm ranges from 365 to 765 nm and 80 wavelengths that are required to enable a deconvolution algorithm of multiple adjacent color centers, the image requires a minimum of 80 acquisitions. If a CCD device has 180,000 pixels (600×300), each file will have a capacity of 14.4 MB (180,000×80).

Future system developments involving high throughput for screening new drugs, medical diagnosis, and cell physiology may require more than 80 data points for 400 nm spectral increments. For an application requiring 200 data points, the file size can grow to at least 36 MB. Massive data acquisition and storage will be required over several minutes before the first computation is

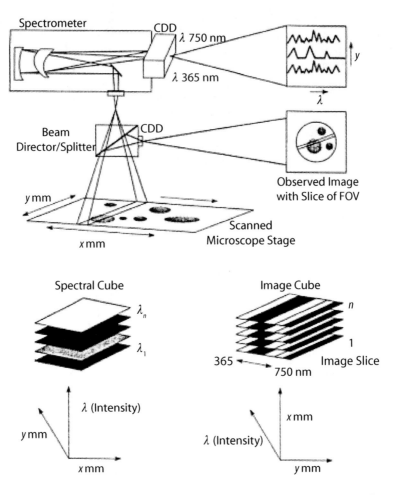

Figure 7-4 *Block diagram showing the critical elements of hyperspectral imaging concept for life science research. Courtesy of John Wiley and Sons, Inc., New York.*

completed. The spectral-cube approach will limit the number of acquisitions, thereby compromising the quality of data obtained.

As stated earlier, the image-cube method captures all wavelength data simultaneously in a 180 K file. An image cube can acquire 240 spectra, each with 750 data specimens in the microscope; visually locate it; target a feature of special interest; and then simultaneously acquire full spectral evaluation of the sample under investigation. This method has potential applications to epitope tagging, cell smears, microtiter wells, and high-throughput scenes. Armed with new imaging methods and sophisticated software technology, researchers will be able to identify the spectral morphology of cells, examine spectral characteristics of cells to indicate a specific disease, and identify appropriate treatment.

7.4 Near-IR Spectroscopic Technique for Epilepsy Treatment

The latest laser technology coupled with state-of-the art FO technology is playing a key role in the treatment of epilepsy. The latest medical research has identified a laser-based technique for pinpointing the location of an epileptic focus in the human brain. The research studies further indicate that the intensity-modulated energy from a near-IR laser diode operating either at a 780 or 830 nm wavelength passes through the scalp, and the reflected energy from the cerebrum is collected to determine the location of the epileptic focus. The reflected beam is collected, detected, and processed to produce a real-time sample of the blood flow to the brain during an epileptic attack. The rate of blood flow and the motion of the flow are critical in the detection of the epileptic focus. This laser-based spectroscopic technique involves noninvasive optical tomography (OT) to obtain 2-D images of the brain surfaces, leading to rapid detection of the brain activity under epileptic seizure conditions [7].

The current method of determining the location of an epileptic focus uses a surgical procedure to allow the measurements of discharges during the seizures, which involves significant cost and great risk to the patient. The computer-based tomography (CT) method involves the injection of a radioactive isotope into the patient's blood to determine the blood-flow increase during the epileptic seizures. Both these methods subject patients to considerable stress, discomfort, and radiation hazard. However, the intensity-modulated near-IR (IMNIR) spectroscopic technique is considered more effective and safe because this technique provides the differences in the absorption index of hemoglobin, measures the changes in the blood flow on both sides of the brain during epileptic seizures, and displays reliable information to neurosurgeons in real time so they can take appropriate action.

A multichannel version of IMNIR spectroscope provides more detailed information on brain activity to the neurosurgeons. A multichannel version of this sensor generally uses eight channels and two diode lasers emitting at 780 and 830 nm wavelengths. Such a system is capable of determining whether an epileptic focus is on the right side or left side of the brain. This sensor permits the location of the epileptic focus within the brain hemisphere with high accuracy and reliability. The ability of the IMNIR spectroscope to obtain time-based variations in the cerebrum blood flow during the epileptic seizure offers an important and reliable diagnostic tool to the neurosurgeons.

7.5 Application of Laser and FO Technologies to Photodynamic Therapy (PDT)

PDT uses laser and FO technologies for promising and effective medical treatment of various diseases such as heart disease, breast cancer, bladder cancer, lung cancer, skin cancer, and deep malignant tumors. PDT uses a photosensitive drug that, when exposed to a specific laser light, creates a toxic form of oxygen known as singlet oxygen. Since the photosensitive drug remains in cancerous tissue for a longer duration than in healthier tissue, the PDT treatment destroys the cancerous tissue selectively. The photosensitive drug is administered to the patient intravenously and collects in the body tissues. When the drug is cleared out of the healthy tissue after a day or two, a surgeon uses an appropriate laser source to activate the administered drug, leading to the destruction of the tissue that is responsible for the cancer growth.

7.5.1 Benefits and Risks Associated with PDT Technology

Unlike chemotherapy, which can make patients sick, the PDT's principal side effect is that the patient's skin may be sensitive to sunlight for a month or so after the treatment. However, PDT has a major limitation—the surgeon must have clear access to the tissue in order to focus the laser beam directly on the area involved. Furthermore, a PDT treatment will be difficult to implement on very large tumors buried deep inside the body due to the unavailability of the tissue.

7.5.2 Potential Illumination Techniques for PDT Treatment

Various laser illumination techniques are available for use in PDT treatment. Clinical scientists are exploring an exciting illumination technique known as laser multiphoto excitation. This technique uses two or more photons from a longer-wavelength laser source that operates in the near-IR region to simulate the same reaction as would be generated from one photon from a short-wavelength laser source. The use of high-power laser sources is absolutely necessary to ensure that multiple photons strike a molecule almost simultaneously. Multiphase excitations offer both spatial control and greater tissue penetration, thereby making this excitation ideal for the treatment of large tumors. With two-photon excitation, some photosensitive molecules can be activated at wavelengths ranging from 600 to 800 nm. Experienced surgeons state that the multiphase excitation in the 740-to-920 nm spectral range causes minimum tissue damage compared to short-wavelength laser sources operating at 300 to 400 nm wavelengths.

Some research scientists have suggested xenon arc lamps and dye lasers pumped either by an argon-ion or frequency-doubled Nd:YAG laser source. High performance LEDs and diode lasers are beginning to replace traditional light sources because of lower cost, higher reliability, and lower power consumption.

Clinical tests to date indicate that LEDs are likely to play a significant role in providing cost-effective light sources for emerging cancer therapies involving PDT technology. Clinical scientists and cancer specialists indicate that PDT can be effective in treating cancers at early stages.

7.5.3 Laser-Source Requirements for PDT Treatment

Laser-source requirements are dependent on the properties of the photosensitive drugs used, type and size of tumor involved, and tissue penetration. Some photosensitive drugs actuate at 632 nm wavelength and some at 652 nm wavelength. It is possible to use KTP-based frequency-doubled Nd:YAG pumped dye lasers or argon-ion pumped dye lasers, depending on the tumor penetration depth and duration of PDT treatment. However, the dye lasers—despite their high power capability—are bulky, consume huge amounts of electrical power, contain toxic dyes, cost over $100,000, and need frequent maintenance. Drug companies are aggressively investigating other drugs, which can be actuated at longer wavelengths ranging from 740 to 920 nm without compromising drug sensitivity.

Solid-state diode lasers and diode-pumped solid-state (DPSS) laser sources are being evaluated for PDT applications with emphasis on cost, drug effectiveness, and reliability. These laser sources are much smaller, more reliable, and relatively less expensive. A solid-state laser AlGaInP diode laser delivers between 1 and 2 W of average output power to the fiber port and works well with some photosensitive drugs. However, a laser with output power greater than 3 W at the end of the output fiber offers excellent results for PDT applications. A new semiconductor laser consisting of a AlGaInP with a strained quantum well (SQW) active region sandwiched between the two AlGaInP waveguide regions, involving AlInP cladding layers is best suited for PDT applications. A monolithic laser array consisting of several SQW devices has demonstrated a CW power level as high as 15 W at 630 nm with maximum power-conversion efficiency of 25%. Such lasers provide power levels of 10 to 15 W at 630 nm using built-in feedback circuits to control the laser-beam intensity on the cancerous tissue. This type of laser design will be most suitable for illuminating the next generation of photosensitizers. Some clinical scientists are using LED devices with longer wavelengths (750 nm or so) to activate specific photosensitive drugs in treating metastatic cutaneous breast tumors [8]. Other researchers are using 732 nm LED devices to test the ability of a special drug to destroy the plagues that contribute to heart disease. Clinical scientists indicate that longer wavelengths are best suited for blood-related treatments because the blood has a low absorption rate at 732 nm wavelength. However, some scientists have selected a blue fluorescent lamp to activate a specific photosensitive drug in treating acute actinic kurtosis because this particular light source does not require precise dosage.

7.5.4 Red LEDs for PDT Applications

Since longer wavelengths are most suited for blood-related treatments because of low absorption characteristics, red diode lasers using AlGaInP/GaAs SQW devices operating at 732 nm are attractive for PDT applications. These diodes use compressively strained structures by adding excess indium content to reduce the laser threshold current. An array of red diodes have demonstrated CW power levels in the 50-to-100 W range with high conversion efficiencies. One of the most impressive applications of such diodes is in PDT treatment of cancerous tumors. A patient is administered a photosensitive drug that concentrates in the tumor after a few hours. The cancerous tumor is exposed to red light at 732 nm, thereby killing all the diseased tissues. Clinical tests indicate that deployment of red-diode lasers for PDT applications makes it possible to use a wide range of photosensitive drugs with maximum effectiveness.

A narrow laser bandwidth of 2 to 3 nm does not take full advantage of the PDT drug-absorption characteristics. According to clinical scientists, the longer the absorption of a photosensitive drug, the more effective the PDT treatment. LED-based lasers operating at 732 or 688 nm offer wide spectral bandwidths in the order of 30 to 40 nm, which will lead to increased laser-energy absorption by the photosensitive drug. Photosensitive drugs exhibit peak absorption at wavelengths over the 670-to-690 nm spectral range, resulting in deeper penetration of the laser light to the affected tumor area. An FO probe with a variable LED wavelength capable of matching the absorption wavelengths of the second-generation PDT drugs will be most suitable for the PDT treatment. Lower cost and wider bandwidth of the LED devices are likely to a play a critical role in the design and development of new light-based equipment for cancer therapies.

7.6 Optical Tomography Using FO Technology for Medical Treatments

Near-IR laser technology is reliable and effective for medical diagnostic and biological imaging applications. This particular technology is a leading contender in the medical-imaging modalities that include magnetic resonance imaging (MRI), CT, and OT. The near-IR laser technology provides improved biological details of specific tissues or organs based on patterns of scattered light at an optimum wavelength. Compared to existing medical and biological imaging technologies, OT offers distinct advantages such as improved reliability, portability, and safety because it does not use ionizing radiation. Portability and reliability features are due to the fact that the laser-based OT equipment does not require large magnets, complicated radiation shields, or bulky power supplies. In addition, OT technology offers therapeutic treatment such as PDT. In the case of an OT procedure, the laser beam illuminates an object at some point on the surface, the laser beam scatters at the bulk material, and an array of sensitive detectors measures the intensity level of the scattered light. The scattered patterns are measured by the detectors, which provide a transmission function of the object for a given source-detector layout. The ultimate solution for OT imaging requires the identification of the internal optical parameters of the object based on the measured transmission function between the laser source and the detector array.

7.6.1 Critical Performance Parameters of the OT Imaging System

Spatial resolution is the most important performance parameter of an OT imaging system, which depends on the number of source-detector pairs in the bulk material [9]. The higher the number of such pairs, the higher the spatial resolution of the system is. High resolution can be achieved by scanning the laser beam over the object by moving the source across the surface. Another approach to achieve higher spatial resolution is to use prior information about the optical properties of the object using an MRI system. However, the reliability of this approach is questionable because of continuous physiological changes in the biological specimen, such as heartbeat and breathing. OT imaging technology is also dependent on the tissue density. It is based on certain assumptions that the medium has weak scattering and absorption coefficients, the distance between two scatters is large compared to source wavelength, and the dielectric constant of the tissue undergoes slow changes. Therefore, in optical atmospheric imaging, the scattering of the signals is independent and can be treated within the framework of transport or diffusion theory, which usually renders excellent results.

The key condition of transport theory is the independence of scattering in a multiscattering system [9], where perturbation of the background medium is relatively weak. At optical wavelengths in the human body, the system breaks down because the distance between the two scatters in a biological object presents the same order of magnitude as the wavelength of the near-IR region around 990 nm. In addition, the local deviations of the dielectric constant are large, so the scattering cannot be considered as independent. Biological tissues have pronounced microstructures—including biological cells and their organelles, blood vessels, and skin and bone structures—which cannot be treated as uncorrelated. Despite these problems, transport theory remains the most promising approach to the problem of optical imaging of biological objects. Consistent good results with real biological tissues are more difficult to obtain because of weak absorption and strong scattering in near-IR region. This means more accurate methods need to be investigated before the OT technology will be able to compete with other techniques such as MRI or CT. The light scattering in biological objects is determined by the dielectric constants of the inhomogeneities on different scales (from microns to centimeters, with an increase by four orders of magnitude), which will require a tomographic imaging algorithm that takes into account heterogeneities on these scales. This algorithm will compute the transmissibility and intensity maps, which are dependent on the number of periodic barriers. Green functions with eigenvalues are used in tomographic imaging to calculate the transmission coefficient of light. Further improvements in OT images and spatial resolution are possible using time gating of the signal by estimating the time resolution for separate spatial trajectories. However, spatial resolution is strictly dependent on the time resolution.

7.7 Endoscopic Sensor Using FO and Laser Technologies

Laser-based endoscopic technology plays a key role in the detection of colon cancers and other rectal disorders. Clinical scientists suggest that an erbium-doped YAG (Er:YAG) laser emitting around 3 microns has potential application for a variety of medical diagnostic procedures, including colonology, dermatology, stomatology, and ophthalmology. Laser wavelengths around 3 microns are attractive for medical diagnosis because human tissue absorbs most of the laser energy at this wavelength, which makes it possible to ablate both the hard and soft tissues. Standard optical fibers are not suitable to handle this wavelength. Optical fibers must be made from materials that transmit signals at 3 microns with minimum insertion loss. FO-based catheters are used for the transmission of such laser signals to tissue for medical diagnosis or treatment.

A practical alternative to the 3-micron laser system involves an optically pumped laser at the other end of the fiber that converts the laser light at near-IR wavelength to a handheld 2.91-micron light source very close to the tissue under treatment. An optical fiber carries the energy from a near-IR Nd:YAG laser to the 2.91-micron handheld laser. This converter laser unit can be miniaturized to a package 2 mm in diameter and 20 mm in length, which is longer than a fiber tip and small enough to be inserted into an endoscope. The device can be cooled by water or air, depending on the heat dissipation and removal rates requirement.

It is possible to employ an Nd:YLF laser operating at 1047 nm to pump ytterbium ions, which can then transfer the laser energy to holmium ions. A more efficient method based on direct pumping of the holmium ions involves pumping with a Nd:YAG laser at 1120 nm, which directly excites these ions into the upper laser level at 2.91 microns. This pumping scheme offers a miniaturized 200 mJ laser source with conversion efficiency better than 25%. This particular laser provides constant energy from room temperature to 100° C, requires no air cooling, and is relatively inexpensive compared to other lasers.

7.8 FO-Based Procedures for Treating Heart Diseases

Transmyocardial Revascularization (TMR) is a laser-based surgical procedure in which tiny holes are drilled into heart tissue for improving blood flow to oxygen-starved heart tissue, thereby relieving the patient of severe angina and other symptoms of severe coronary disease. An FO catheter is used to deliver the laser energy where it is needed the most with maximum efficiency. Holmium-based YAG (Ho:YAG) lasers and thulium-based YAG (Tm:YAG) lasers that emit around 2 microns are best suited for TMR procedures. This eye-safe wavelength does not reach the retina. The 2-micron wavelength is strongly absorbed by water and affects only the outer layers of the tissues. This wavelength laser energy can be easily transmitted or delivered via silica fibers with minimum insertion loss. Critical elements of the Ho:YAG laser are shown in Figure 7–2. The Ho-base for a TMR procedure is relatively costly, less traumatic to patients, and requires no blood transfusion or heart-lung equipment. After this procedure, recovery is faster, more cost-effective, and freer from postoperative complications than open-heart surgical procedures [10].

TMR may be considered as a replacement therapy to bypass surgery or angioplasty and could become the third major medical advancement for the treatment of heart disease in the near future. In the case of laser-based angioplasty procedures (see Figure 7–5), the deposits in the artery can be melted and removed from the artery walls using the inflated balloon. According to clinical researchers, the CW laser power requirement of 15 W seems reasonable, the inflation pressure must be kept below 30 psi, the duration of laser heating must not exceed 50 seconds, and the tissue temperature must be monitored constantly during the procedure for the safety and comfort of the patient. This laser-based angioplasty procedure is still under clinical investigation.

Clinical scientists indicate that a percutaneous (procedure performed through the skin) TMR procedure can be relatively less complex. This procedure relies on FO catheters to deliver the laser energy to the heart and can be characterized as a minimally invasive procedure. Clinical evidence of the safety and the effectiveness of the percutaneous approach suggest that the medical community would prefer percutaneous myocardial revascularization (PMR) compared to the TMR procedure.

In a PMR procedure, the surgeon places the laser probe against the ventricle wall for drilling the holes in the myocardium. Furthermore, a PMR procedure requires only a few holes (10 to 20) compared to a TMR procedure, and the entire procedure takes only one to two hours to complete. The entire PMR procedure can be performed under local anesthesia in a cardiac-catheterization laboratory rather than in a surgical suite, and the patient goes home in a day, compared to three or four days following a TMR procedure.

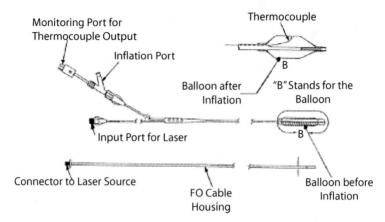

Critical Element Description:
- IR laser to melt the plaque.
- Balloon inflated to remove the melted plaque from the artery.
- Thermocouple to monitor the temperature of the balloon surface.

Figure 7–5 *Laser-based angioplasty procedure to clear blocked heart arteries.*

7.8.1 Deployment of Stent Technology for Clearing the Arteries

The latest laser-based machine technology plays a critical role in the development of stainless-steel stents widely used to clear heart arteries with excellent results. High-power lasers are used to machine minute precision parts from surgical materials for implantable devices such a stents, catheters, and needles. One of the most commonly applied implants is the coronary stent made from stainless steel. This hollow tubular mesh is used following balloon angioplasty to prop open an artery and to prevent restenosis, or recurrence of narrowing of the artery. Restenosis occurs when the scar tissue that forms as a part of the healing process becomes too thick and obstructs the cleared artery again. According to clinical cardiologists, the restenosis condition can occur in 20% to 30% of patients. In the case of diabetic patients, this rate can exceed 50%. The balloon is inflated to place the stent, the tiny meshlike tube, at the desired location in the artery. The vessel-narrowing plaque or tissue is held in place by the stent. The microstrucure of the stent is a delicate scaffold composed of high-quality stainless steel or a platinum alloy, which allows the artery to grow in and around the material until it becomes a part of the arterial wall. The geometry of this structure is critical to how it will function once it has been implanted in the artery. High-precision machining is critical in the design of stents.

An automatic drug-delivery mechanism is given serious consideration in the design and development of future stents. The stents can be bonded with appropriate drugs that will release medicine into the surrounding tissue to prevent restenosis. A polymer coating covering the whole stent will release the drugs into the vessel wall. A restenosis rate well below 5% can be expected with advanced stents.

7.9 Vision Correction Using Laser Surgical Procedure and FO Technology

Laser surgical procedures have been very successful in 20/20 vision correction and treatment of other eye diseases over the last two decades. The laser-based photorefractive keratectomy (PRK) technique is effective in treating nearsighted patients. In a PRK procedure, the outer layer of the cornea is polished off and then the laser beam is used to reshape the surface of the cornea. The entire procedure is done at the doctor's office under local anesthesia, and the patient goes home on the same day. The patient may experience some blurry vision and discomfort for a couple of days.

The latest laser-based technique, called laser in situ keratomileusis (LASIK), allows the surgeon to make a very thin flap in the outer layer of the cornea, known as the epithelium [11] and to lift this up while the inside of the cornea is under laser treatment for a few seconds. The flap is then put back in place on the epithelium. The LASIK technique allows the patient to see well the next day without any discomfort. The outcome of LASIK surgery is dramatic, and patients who have undergone such procedure like it very much. Surgical data and comments from patients indicate that about 70% of the patients have regained 20/20 vision without glasses or contact lenses and 90% have achieved 20/40 vision.

7.9.1 Laser Requirements for Diagnosis and Treatment of Eye Diseases

Laser-based optical coherent tomography (OCT) technology was originally developed for ophthalmology to provide 10-micron resolution for imaging the retina surface. OCT is very effective for the diagnosis and monitoring of eye diseases such as glaucoma and macular edema in diabetic retinopathy because it can measure the progress of the disease with great reliability.

Advanced solid-state lasers emitting femtosecond pulses and capable of operating over a wide range of wavelengths are available for eye treatments and LASIK surgical procedures. Such ultranarrow-pulse lasers include Nd:YAG, Nd:YLF, and Cr:YAG solid-state lasers. A low-cost, diode-pumped Nd:glass laser with hundreds of femtosecond (fs) pulses is best suited for glaucoma treatment because the duration of hundreds of fetmoseconds allows the laser energy to get in and out very quickly and the ablation threshold becomes the deterministic parameter. Furthermore, with this femtosecond-pulse format, laser energy can be adjusted to avoid collateral damage. This is a critical requirement in all laser medical procedures but particularly so when treating the eyes. The shorter the laser pulse, the broader the optical bandwidth, which is an important factor in laser surgical procedures and WDM communications systems. Squeezing a pulse in time concentrates its energy in time and focuses the laser energy on the area of interest. Short pulses provide high peak power levels, and extreme precautions must be taken to avoid damage to the surrounding tissues. Laser pulses as short as 5 fs have been obtained by compressing the 13 fs pulses from a Ti:sapphire laser by first extending its bandwidth from 100 to 500 nm, then compressing their duration to less than 5 fs.

7.9.2 Short-Pulse Lasers for Various Medical Treatments

Short-pulse lasers are widely used for cancer treatment, radical keratotomy, and laser-based angioplasty. Short-pulse excimers have joined other lasers in gaining wide acceptance in medical diagnosis, surgical procedures, and testing of chemical compounds. Raman spectroscopy using short laser pulses offers early detection and diagnosis of certain life-threatening diseases. Ultrashort laser (USL) pulses cause minimum collateral damage and thermal damage to the surrounding tissues because of very short energy deposition time and high efficiency of the ablation process. Minimum energy deposition into secondary tissues means less discomfort to the patient. The USL could be of significant importance in sensitive and risky procedures, such as spinal surgery, where collateral damage could have devastating consequences for the patient. USL pulses are being investigated for their ability to improve pain management and relief from angina pain. Studies performed by the author on various lasers indicate that an Nd:YAG laser with CW output of 36 W at 500 nm and 100 W at 1060 nm, a holmium (Ho) laser with CW output of 30 W at 2100 nm, and a thulium (Tm) laser with CW output of more than 20 W at 1960 nm are available for various surgical procedures.

Excimer lasers are investigated for various medical treatments. Ecimer lasers using rare gas halides such as ArF (193 nm), KrF (248 nm), XeF (353 nm), KrCl (222 nm), and XeCl (308 nm) are being considered for various medical and surgical applications. An XeCl laser emitting at 308 nm, when coupled through an UV fiber bundle, is best suited for angioplasty to clear a blocked human artery. Longer-duration pulses are recommended to propagate maximum laser energy per pulse to the arteries for a rapid clearing process with minimum discomfort to the patient. An excimer laser involving ArF shortpulses at 193 nm permits tissues to be cut cleanly and with great precision. Hence, it is best suited for radical keratotomy procedures or direct reshaping of the lens curvature. Because of the wide bandwidth capabilities of the excimer lasers, they are excellent candidates for generating ultrashort pulses and high-power laser subpicosecond pulses, which can be amplified to terrawatt-level outputs by employing discharge power amplifiers.

7.10 Laser Technology for OT Applications

Near-IR laser technology is considered to be the most reliable and effective tool for medical and biological imaging and could be a leading contender in medical imaging modalities, which include MRI, CT, and OT. The near-IR laser technology provides improved biological details of specific tissues or organs based on patterns of scattered light at optimum wavelength. Compared to existing biological imaging technologies, OT offers distinct advantages, such as enhanced reliability, higher resolution, portability, and safety because it uses no ionizing radiation. OT is reliable and portable because it does not require large magnets, radiation shields, or huge power supplies. OT imaging technology offers therapeutic treatment only through the time gating of the signal by estimating time resolution for separate spatial trajectories. However, the spatial resolution is dependent on time resolution.

7.10.1 Performance Requirements of OT Imaging Systems

Spatial resolution is the most important performance parameter of an OT imaging system and is dependent on the number of source-detector pairs in the bulk material. The higher the number of such pairs, the higher the spatial resolution is. This can be achieved by scanning the laser beam over the object or moving the optical source across the surface of the object. Another approach to achieve higher spatial resolution is to use prior information on the optical properties of the object from an MRI system. However, the reliability of such an approach is questionable because of continuous physiological changes such as heartbeat and breathing. OT technology essentially involves tissue density. It is based on certain assumptions that the medium has weak scattering and absorption coefficients, the distance between the two scatters is large compared to the operating wavelength, and the dielectric constant of the tissue undergoes slow change. The light scattering in biological objects is determined by the dielectric constants of inhomogeneities on different scales, which will require a tomographic imaging algorithm that takes into account heterogeneities on these scales. Sometimes Green functions with eigenvalues are used to obtain tomographic imaging to compute the transmission coefficient of the light.

7.11 Q-Switched Laser for Clearing a Cerebral Artery Obstruction

Q-switched laser pulse energy at 532 nm, when delivered through a novel FO catheter, plays an important role in eliminating the blood clot associated with an ischemic stroke [12]. The specially designed system uses an FO catheter to deliver the 532 nm light energy from a frequency-doubled, Q-switched Nd:YAG laser. The laser energy is focused on the blood clots lodged in a cerebral artery to eliminate the clots, thereby saving the patient's life. This laser procedure requires a surgeon very familiar with the performance parameters of the 532 nm Q-switched laser. Precision focusing, rapid heat generation, and immediate cooling are involved in this procedure.

The 532 nm, Q-switched Nd:YAG laser with average power close to 5 W and operating at 5 kHz emulsifies the blood clot by cavitation (a phenomenon involving a formation of partial vacuums in a liquid by a fast-moving solid body) due to rapid heating and cooling caused by the laser pulses. Not only for its simplicity and reliability, but also the 532 nm laser is preferred because it results in maximum differential absorption between the blood clot and artery walls, thereby minimizing the possibility of damage to the surrounding tissue. Protection of the surrounding tissue is critical in all laser-related treatments. This particular laser is best suited for biomedical and precision surgical applications due to its high reliability and safety to patients. A wide range of laser technologies, including gas, solid-state, and ultrafast lasers are available for various medical and surgical procedures. Each laser source occupies a unique spot in the price/performance continuum.

7.12 Laser-Based Flow Cytometry

Laser-based flow cytometry (a science dealing with the life history of cells) is widely used in medical research and clinical applications needed for sperm sex selection, dairy cattle breeding and sorting fetal cells from maternal blood samples. In flow cytometry, cells are made to pass in single file, either in a liquid stream or in discrete droplets, through a laser interaction zone. The cells are pretreated with fluorescent markers that preferentially adhere to various cell types. The CW laser-based system detects the fluorescence using a series of photodetectors (see Figure 7–6), each of which features its own spectral filter. The resulting fluorescence is separated and detected using one or more photomultiplier tubes, each with a specific optical passband filter. The intensity and/or the spectral composition of the resulting fluorescence allow the cells to be counted, analyzed, and sorted using the electric field [12]. In sorting, a small static charge is applied to the cells so that they can be deflected into receptacles depending on their fluorescent-signal levels.

The laser is a critical component in the flow-cytometry instrumentation system. It allows a large light flux to be coupled into a small area, rendering it capable of producing sufficient fluorescence intensity for instant analysis of single cells. Most of the currently used flow-cytometry instruments use the air-cooled, 488 nm argon-ion lasers because of simple and low cost designs. This instrumentation system, if it incorporates a red HeNe laser or diode laser operating at other wavelengths—such as a UV water-cooled ion laser—is best suited for multiparameter biomedical research studies.

Air-cooled ion lasers are simple, low-cost light sources of blue light with high conversion efficiency. Alternative solid-state 488 nm laser designs that

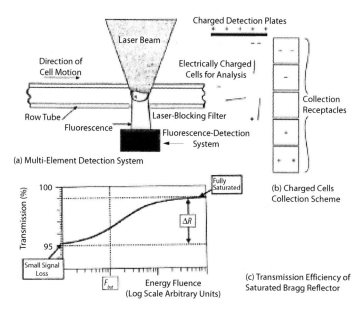

Figure 7-6 *Laser-based flow-cytometry system for biomedical and life-science research studies.*

promise smaller package size and lower power consumption are under development. However, the existing laser source of choice is the large-frame water-cooled ion laser, which suffers from high procurement cost and short operating life. A new alternate to the laser is the frequency-tripled quasicontinuous-wave all-solid-state laser. This particular laser operates in a pulse mode with pulse repetition rates as high as 80 MHz, and acts like a CW laser in the context of cytometry and samples up to 50,000 cells per second. This cell-sampling rate is attractive for biomedical research studies. Based on the advantages and disadvantages associated with various laser sources, one can conclude that a Q-switched laser with high pulse energy is best suited for medical procedures and biomedical research studies. Furthermore, lasers with high peak power levels, picosecond pulse durations, and high repetition rates are preferred for flow cytometry and precision surgical procedures.

7.13 Laser-Based Endoscopic Technology for Colon Imaging

Laser-based endoscopic technology is getting more popular for colonospy applications. An endoscope instrument consisting of a laser source and a SM FO probe eliminates the need for a high-speed scanning element. It also reduces the diameter of the probe and thus provides maximum comfort to the patient. Recent advances in hybrid imaging techniques and microendoscopes have demonstrated a significant improvement in the imaging of structural and functional aspects of pathological processes. But this endoscope hybrid technology is still in development, is very expensive, and could take several years to satisfy the promise of a truly comprehensive, noninvasive diagnostic tool for colon-imaging procedures. Meanwhile, state-of-the art, miniature endoscopes are widely employed for colon procedures [13].

7.13.1 Design Aspects of Miniature Endoscopes

The smallest, state-of-the art endoscope has a probe diameter of less than 1 mm or 0.0394 inches. The miniature endoscope can now image previously inaccessible parts of the body with remarkable resolution. A typical probe consists of bundles of low-loss optical fibers, each bundle conveying one pixel of an image. However, the practical dimensions of the probe, which reduce the FOV and the resolution of the image, limit the number of fibers that can be included in the bundle. Fast acquisition of images requires rapid scanning in one direction to acquire an image, which can be accomplished with a galvanometer or with a device using the MEMS technology. This MEMS device offers scanning speeds best suited for obtaining remarkable images of the colon in the shortest possible time and with minimum discomfort to the patient.

A wider FOV with much greater resolutions is possible with diffraction-grating technology. Volume-phase-diffraction grating technology allows the laser light to spread across the sample. The grating illuminates the sample or specific object area with an array of focused spots, where each position is encoded by a different wavelength. The reflected light is then transmitted back through the optical fiber and is decoded externally. A line of image can be acquired by measuring the reflected spectrum using a 15 kHz spectrometer. Transverse mechanical scanning of the optical fiber and distal optics in the conjugate direction must occur to obtain a 2-D image of the object. High grating-diffraction efficiency is necessary to meet the above requirements.

When a grating device with grating efficiency of better than 90% is coupled to a Ti:sapphire laser with a bandwidth of 225 nm centered at a wavelength of 830 nm, an endoscopic probe can achieve a much wider FOV with greater resolution at 150,000 resolvable points, compared with 30,000 with an optical fiber bundle probe. This particular technique offers five-to-one resolution improvement over a wide FOV. This technique offers an ultraminiature, single-fiber endoscope with a body diameter of 250 microns or 0.010 inches and probe diameter of 500 microns or 0.020 inches, which can image parts of the human body that were previously inaccessible. Deployment of ultraminiature endoscopes will minimize damage to sensitive tissue and bleeding sometimes associated with endoscopy procedures. The ultraminiature endoscope will also reduce the risk of these procedures by decreasing the need of anesthesia in certain cases.

7.14 Fiber-Based Delivery Systems for Medical Applications

Recently developed one-dimensional photonic bandgap fiber delivery technology has potential applications at any wavelength. The omniguide optical fiber exhibits its best transmission characteristics in the region of the primary bandgap at 10.6 microns. In addition, this optical fiber is capable of transmitting other wavelengths in the higher-order bandgap. An optical designer can position the primary bandgap anywhere in the IR spectral range. This particular optical fiber has potential applications in remote sensing, telecommunications, commercial sensors and industrial systems. This fiber will also find wide acceptance in a variety of medical procedures.

Clinical scientists, physicians, and surgeons can use this unique fiber in a variety of medical procedures in fields including ophthalmology, general surgery, dermatology, dentistry, and veterinary medicine. In the case of surgical procedures, the laser light must be delivered from a stationary laser source in the operating room to the surgeon's hand, where the real action is. Optical fibers tend to add a degree of flexibility by simplifying laser-beam handling in traditional surgery as well as facilitating the endoscopic beam delivery. Some of the most useful medical and surgical lasers, such as Er:YAG and CO_2 lasers operating at appropriate wavelengths, preclude the use of the established silica-fiber technology. Under these situations, one-dimensional photonic bandgap optical fibers are ideal.

In addition to medical applications, these fibers could be used in space sensing, scientific research, telecommunications, industrial systems, and military equipment, where conventional fibers cannot be used because they are not able to meet the performance requirements for these specific applications. These fibers can be used for remote, noncontact sensing at mid-IR wavelengths in scientific-research applications such as spectroscopy and high-resolution thermal sensing.

The nonlinear transmission properties and polarization mode dispersion in most solid-core fibers distort optical signals carried by long transmission lines. Cost benefits of hollow-core fiber technology are possible by eliminating the need for amplifying signals every 50 to 100 km and periodically regenerating them. Significant advantages could be expected by using such fibers in submarine networks.

Lasers generating optical power levels greater than 1 kW are used in industrial and commercial applications. The major application is metal processing involving cutting and welding in a variety of industries, including automotive assembly, shipbuilding, and the process of nuclear plant decommissioning. High-power Co_2 lasers are widely used in military, space, and medical applications. In critical applications such as optical missile tracking, target illumination, and towed ECM systems (see Figure 7–7), laser beam delivery with no distortion in pulse shape or beam circularity are critical. In the case of towed ECM decoys, the

FO cable (see Figure 7–7) must provide high mechanical integrity under severe dynamic conditions, in addition to low transmission losses and a reduction in the beam's noncircularity problem. A high-power solid-state Nd:YAG laser (1.064 microns), high-power gaseous CO_2 laser (10.6 microns), and high-power chemical oxygen iodine laser (COIL) (1.315 microns) are best suited for certain industrial, military, and medical applications. Regardless of the application, the FO-based delivery system must provide reliable operation without compromising optical performance and mechanical integrity while operating under severe thermal and mechanical environments. The first two lasers are widely used in cutting, drilling, and welding, whereas the last laser is most suitable for destroying long-range hostile missiles and for removal of debris from the decommissioning of nuclear power plants. Nd:YAG laser technology offers the highest operational flexibility. This flexibility implies that engineers now can mount an optical fiber guiding an Nd:YAG laser beam on a robot, which can be integrated into a highly automated steel-welding system in an automotive-assembling plant or heavy-machinery production plant.

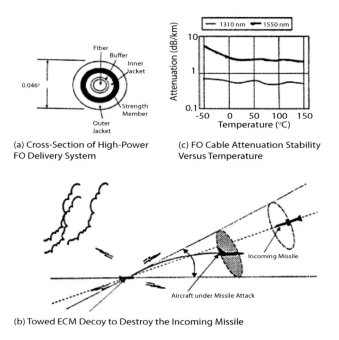

(a) Cross-Section of High-Power FO Delivery System

(c) FO Cable Attenuation Stability Versus Temperature

(b) Towed ECM Decoy to Destroy the Incoming Missile

Figure 7–7 *High-power FO delivery system (a) for the towed electronic countermeasures (ECM) system (b) to destroy the incoming hostile missile. Attenuation stability is important.*

7.15 Summary

Potential applications of FO technology in the areas of medical treatment, scientific research, and life sciences are described in greater detail. The hybrid use of FO technology and advanced laser technologies is investigated for various medical treatments and diagnosis of diseases. Potential applications of these technologies in medical and surgical procedures in ophthalmology, dentistry, general surgery, life science, and veterinary medicine are identified. Major advantages of a laser-based DNA technique, known as the capillary electrophoresic process, used by forensic scientists to identify criminals are summarized. Optical fiber requirements for DNA tests, dental surgery, and high-power laser transmitters are identified with emphasis on reliability, safety, and costeffectiveness. A near-IR spectroscopic technique for epilepsy treatment is described. Benefits of laser-based techniques for PDT, OT or imaging, colonoscopy, angioplasty, and TMR are described with emphasis on safety and comfort of the patient. A Q-switched laser technique using an FO catheter to emulsify the blood clot in the cerebral artery and to prevent an ischemic stroke is discussed, including the identification of the critical fiber and laser requirements. Benefits of CW UV lasers, such as excimer lasers, are discussed for life-science research, bioscience investigation, and precision surgical procedures. Critical requirements of fiber-beam delivery systems for various applications are briefly summarized with emphasis on stable performance under harsh thermal and mechanical environments.

7.16 References

1. Appel, D. (1998, April). Laser application in DNA analysis. *Laser FocusWorld*, 46–47.

2. Jha, A. R. (2000). *Infrared technology: Applications to electro-optics, photonic devices, and sensors* (p. 229). New York: John Wiley and Sons, Inc.

3. Kincade, K. (1998, July). Raman spectroscopy enhances into vivo diagnosis. *Laser Focus World*, 83–97.

4. Jha, A. R. (2000). *Infrared technology: Applications to electro-optics, photonic devices, and sensors* (pp. 291–292). New York: John Wiley and Sons, Inc.

5. Marty, B. R. (1999, February). New research could end dental filling. *Laser Focus World*, 34.

6. Lerner, L. (1998, October). Hyperspectral images for life science vistas. *Laser Focus World*, 89–90.
7. Jha, A. R. (2000). *Infrared technology: Applications to electro-optics, photonic devices, and sensors* (p. 291). New York: John Wiley and Sons, Inc.
8. Robinson, K. (1998, May). Photodynamic therapy offers new medical treatments. *Photonics Spectra*, 219–226.
9. Jha, A. R. (2000). *Infrared technology: Applications to electro-optics, photonic devices, and sensors* (pp. 296–297). New York: John Wiley and Sons, Inc.
10. Jha, A. R. (2000). *Infrared technology: Applications to electro-optics, photonic devices, and sensors* (p. 292). New York: John Wiley and Sons, Inc.
11. Jha, A. R. (2000). *Infrared technology: Applications to electro-optics, photonic devices, and sensors* (p. 299). New York: John Wiley and Sons, Inc.
12. Krueger, A. (2003, May). Q-switched laser for clearing cerebral artery obstruction. *Photonics Spectra*, 82–83.
13. Johnson, B. D. (2003, October). Spectrally-encoded endoscopy technique. *Photonics Spectra*, 38–39.

CHAPTER 8

Fiber Optic Sensors for Various Industrial Applications

This chapter focuses on the FO-based sensors that are best suited for industrial applications. Such sensors include level sensors widely deployed in the petrochemical and chemical industries; temperature sensors for monitoring various industrial processes; displacement sensors in machining precision parts for industrial, military, and medical applications; pressure sensors to monitor absolute pressure or pressure change in industrial processes; intensity-modulated sensors to measure electric or magnetic fields; and interferometer sensors to detect phase shift in the optical or RF signals. All these sensors use fiber optics as transmission lines for transmission and distribution of optical signals with minimum insertion loss and dispersion. Immunity to EMI, RFI, and nuclear radiation makes fiber optics most desirable in many diversified applications such as industrial, military, and space applications. Furthermore, FO technology offers high reliability under severe operating environments.

In all these sensor applications, custom-built optical connectors are required to ensure reliable and stable optical performance. In special cases, a cantilevered fiber and heat-sink chip are incorporated to significantly reduce fiber breakdown, usually the result of beam misalignment in which laser energy is focused outside the fiber core, causing serious fiber damage. A miniature heat sink is attached to the rear of the optical connector to conduct extraneous heat away from the fiber, which can further reduce the possibility of fiber breakdown. In the case of heavy industrial applications, the sensors must use the fibers that are double-sheathed in super-rugged, flexible stainless-steel square-lock cable with an internal PVC/Kevlar sheath to provide improved mechanical integrity under severe mechanical stresses and vibrations. In brief, these sensors require FO cable with low loss and dispersion over a wide spectral bandwidth to provide high optical stability, regardless of the operating environments.

Studies performed by the author on industrial cables indicate that a special 1.2 mm FO cable is available to provide consistent optical performance under severe thermal and environmental conditions and to meet the requirements of today's industrial optical systems. This particular FO cable consists of a pure silica core; a polytetrafluoroethylene (PTFE) buffer layer that effectively decouples the optical fiber from other cable elements; a fluoropolymer inner jacket for added strength and isolation; a Kevlar aramid strength member to provide additional strength; and a steel extruded jacket to guard against an accidental cut, bruise, or bend. This cable has demonstrated stable optical performance when subjected to vibration, mechanical shock, tension, and temperature extremes. One can expect such operating environments in most industrial applications.

8.1 FO Level Sensors

Level sensors can be categorized as switches for high/low levels, leak detection devices, or magnitude sensors for monitoring fluid levels in a tank. FO devices have found the most applications in the level sensor category. Liquid level is one of the most important process control parameters and is widely used in the petrochemical and chemical industries. The explosive nature of many of the processes makes FO attractive. Leak detection and fuel-level measurements are required in many industrial and military applications. The immunity to EMI and RFI is the driving force for the deployment of FO technology in the above applications. Furthermore, FO sensors that use light interaction with the fluid for which the level is being measured work best in relatively clean and clear liquids. Dirt or opaque liquids, such as crude oil and paints, tend to foul optics and blind the sensor, as do solids in powder form. The operating principal of FO level sensors can be based on several concepts [1], such as sight glass, force or buoyancy, pressure, reflective surface, and refractive index.

8.1.1 Sensor Design Based on Sight-Glass Concept

This type of sensor design is best suited for pressure measurements in thermal power plants, where precision levels of saturated steam, superheated steam, and hot water in boilers need to be monitored or observed frequently from a safe distance. Steam and hot-water levels are frequently monitored to maintain high entropy essential for high steam-turbine efficiency. Any moisture content in the steam will reduce turbine efficiency and deteriorate the high-speed turbine blades.

A predetermined water level in a boiler is required all the time to make available the desired supply of steam to the turbine. As the water or liquid interrupts the light-source path, the water level is determined by measuring the transmitting coefficient of the signal, which is calibrated in inches of water from a predetermined level.

8.1.1.1 Operating Principle of the Sight-Glass Sensor

High-pressure steam boilers use a bicolor visual sight gauge for water level. A prism is used in such a way that a red light comes on if steam is in the gauge port. Water in the port causes the red light to be refracted, with only green light transmitted. Such a device concept is illustrated in Figure 8–1.

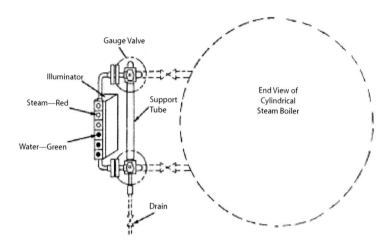

Figure 8-1 *Bicolor liquid-level sensor showing the steam and water levels in the cylindrical steam boiler.*

The location of the sensing device has to be at the boiler site, which is normally inconvenient for central monitoring. If the sensor is equipped with a large core, i.e., a large-NA optical fiber, then the green/red water/steam signal can be transmitted to a central monitoring area as shown in Figure 8-2. The interface of the fibers and color ports occurs within a hood to minimize any ambient-light interference. The sensor is completely passive, with no moving or electrical components other than the illuminator. This approach is required by many existing safety codes. Since the level of the transmitted light must be high enough to be visible, the distance to the remote monitoring location is limited to about 100 meters. Critical components for a transmission-type level sensor are shown in Figure 8-3. Liquid level is visible through the sight glass.

8.1.2 Force-Type Liquid-Level Sensor

In a force- or buoyancy-type level sensor, a float is generally used to track the liquid level. A simple float device, such as the one shown in Figure 8-4(a), is required to actuate a reflective FO sensor when a reflective target attached to the float passes the probe. A similar device shown in Figure 8-4(b) uses a transmission-type FO sensor. Sensors of this type are generally used for high- or low-level detection only. If the interruption device is designed with a transmission binary code plate and multiple transmission-sensing fibers, the actual height of the liquid can be determined in a digital format. With an 8-bit code, the liquid level can be resolved to one part in 256.

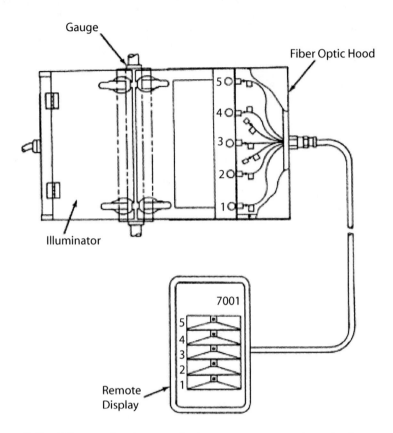

Figure 8–2 *FO water-level gauge viewing system at central monitoring location 100 meters away from operating equipment. Courtesy of Instrument Society of America, Research Triangle, North Carolina 27709.*

8.1.3 Liquid-Level Sensor Using Pressure Transducer

Liquid level can be determined by measuring the hydrostatic head, the pressure exerted by the level of the liquid. A mathematical expression relating level height H and pressure P can be written as

$$H = [P/\rho\delta] \qquad 8.1$$

where ρ is the density of water and δ is the specific gravity of the liquid. Using various constants involved, one will find a liquid level of 30 feet of water ($\delta = 1$) equals a pressure of 13 psi, 30 feet of oil ($\delta = 0.65$) equals 8.44 psi, and 30 feet of brine solution ($\delta = 1.3$) equals 16.9 psi. Studies performed by the author indi-

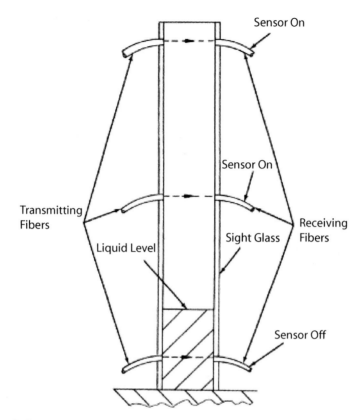

Figure 8–3 *Transmissive liquid-level sensor using sight glass.*

cate that due to higher pressure levels, high-density liquids have a higher degree of accuracy in fluid-level measurements than low-density liquids.

A reflective FO sensor with a diaphragm is widely used to obtain liquid-level measurements. The critical elements of this type of sensor are shown in Figure 8–5. Typical water-tank levels range from 20 to 60 feet, with a pressure ranging from 0 to 26 psi. Resolution of 6 inches corresponds to a pressure-sensor accuracy of +/- 0.21 psi or +/-1%, approximately.

The pressure measurement is complicated if the tank is pressurized. The measurement system requires two pressure sensors: one at the top of the tank to measure pressurization and one at the bottom of the tank to measure the hydrostatic head plus pressurization. The difference in the pressure values corresponds to the liquid level in the tank. If pressurization is high, small errors in pressure measurement can cause large errors in liquid level [1].

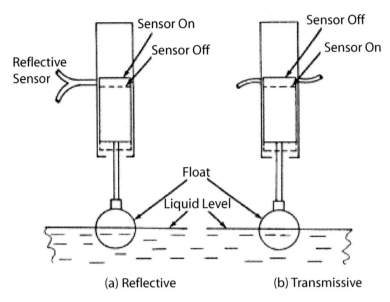

Figure 8-4 *Force- or buoyancy-type liquid-level sensors using (a) reflective and (b) transmissive design concepts. Courtesy of Instrument Society of America, Research Triangle, North Carolina 27709.*

8.1.4 Liquid-Level Sensor Based on Surface-Reflectance Technique

The surface-reflectance technique uses the reflectance of light from the liquid surface to determine the liquid level in a tank or container [1]. For an incident light striking the liquid surface, the position of the detected beam is strictly a function of liquid level. It is relatively straightforward to use optical fibers to locate electronic and optical components (light source and photodetector) away from the measurement area.

A large dynamic range or very high sensitivity would require a large number of detector elements. This approach is particularly attractive for corrosive liquids or high temperature liquids where physical contact is not possible. For example, such an approach would be best suited to measure the level of molten glass. If the light source is modulated (pulsed) by the IR source, the pulsed signal and not the ambient light from the molten glass will be detected. Molten glass also tends to be highly viscous and have a smooth surface, making surface reflectance more effective. Low-viscosity liquids are subject to surface vibrations causing ripples, so they are not good candidates for this approach.

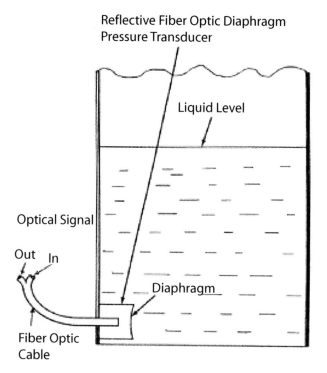

Figure 8–5 *Liquid-level sensor using FO pressure transducer.*

8.1.5 Liquid-Level Sensors Based on Refractive-Index Change

The operating principle of a liquid-level sensor based on refractive-index change requires transmission of light to a prism, typically a quartz prism whose refractive index is 1.46. In the medium of air (refractive index = 1.0), the prism acts as an FO with the air being the interface medium. Since the refractive-index-change sensor requires small amounts of liquid to switch, leaks at lubrication seals can be easily detected. This technique can be used in the detection of air leaks in a pipe or, if attached to an electronic counter, the determination of the smallest fraction of liquid.

Liquid level for a specific liquid can be determined by measuring the reflected light signals coming from the ground prism after the liquid surface is illuminated by the UV (300 to 800 nm) laser source through the FO rod as illustrated in Figure 8–6. The output of the sensor for various liquids can be determined by the ratio of FO-to-prism refractive-index change. The relative output of the sensor for various liquids using the FO-to-prism refractive-index change is summarized in Table 8–1.

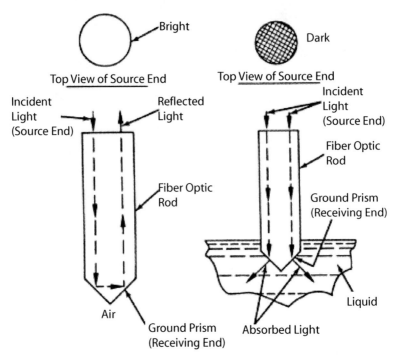

Figure 8–6 *FO liquid-level sensor based on refractive-index change (reprinted with permission from SENSORS, 1986, published by Helmers Publishing, Inc., Figure 10.10).*

8.2 FO-Based Displacement Sensors

FO displacement sensors play a major role in industrial, military, and medical applications. Two major advantages of this type of sensor include the potential for the most accurate noncontact sensing and the possibility of incorporating the optical sensors permanently in the composite structures. The basic FO intensity-modulated sensor and an interferometer-type sensor are capable of measuring displacement parameters. These two sensor designs have provided the primary approach to displacement sensors: reflective and microbending sensors. The following paragraphs will focus on a detailed review of how these sensors function to meet displacement-sensor performance requirements in various industrial applications.

8.2.1 Displacement Sensor Using Reflective Technology

The basic concept of a reflective sensor is depicted in Figure 8–7. The sensor consists of two FO legs. One leg transmits light to a reflective target; the other leg

Table 8–1 *Sensor output for various liquids using refractive-index change.*

Medium	Relative Sensor Output
Air	1.00
Water	0.11
Isopropyl alcohol	0.07
Gasoline	0.03
Milk	0.20

traps reflected light and transmits it to a sensitive detector. The intensity of the detected light is dependent on the location of the reflecting target from the FO probe. The sensor response curve shows a maximum with a steep linear front slope and a back slope, which follows $1/R^2$, where R is the distance between the tip of the FO probe and the reflecting surface. The response-curve characteristic is based on geometric optics. Light exits the transmitting fibers in a solid cone confined by the NA of the core. The spot size hitting the target can be expressed as

$$D = [2R \tan \theta] \qquad 8.2$$

where θ is the half angle between the normal to the fiber exit surface and the exit divergence cone (NA) and D is the spot size or diameter. Since the angle of reflection is equal to the angle of incidence, the spot size that impinges back on the FO probe after the reflection is twice the size of the spot that hits the target initially. As the distance from the reflecting surface increases, the area of the spot increases and is directly proportional to R^2. The amount of the detecting light is inversely proportional to the spot area, or $1/R^2$. As the probe tip comes closer to the reflecting target, there is a position in which the reflected light rays are not coupled into the receiving fiber.

Employing appropriate values of optical parameters and correct sensor configurations, one can optimize the sensor output. The dynamic range and sensitivity can be tailored for specific application. Highest sensitivity occurs at the closest target distance and vice versa. The fiber pair probe has the largest transmit fiber to receive fiber spacing (center-to-center) and, therefore, is least sensitive at close target distances. In a single-fiber sensor configuration, the front slope disappears because when the reflecting surface is approached, light continues to be reflected into the same fiber, which is functioning in both transmit and receive modes.

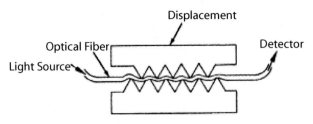

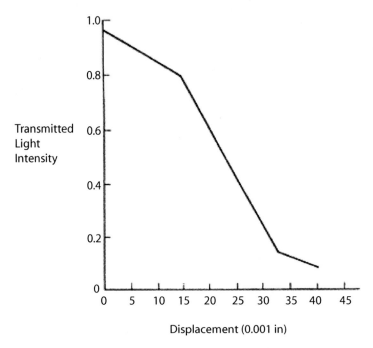

(a) Illustration Showing the Microbending Concept

(b) Transmitted Light Intensity as a Function of Displacement

Figure 8–7 *FO displacement sensor using microbending.*

A dual-probe scheme improves sensor sensitivity significantly. The probes are spaced so that the reflected intensity is the same for the probe in front of the slope as in back of the slope. As the target moves closer, the detected intensity of one of the probes decreases, while the detected intensity for the other probe increases. A dual-probe scheme not only increases the sensor sensitivity, but also provides displacement direction information and magnitude. The highest sensitivity is on the front slope of the trough. Both the dynamic range and sensitivity can be adjusted by arranging the focal length of the focusing lens. Movement can be detected with an accuracy of 0.001 inches over a dynamic range of 1 inch using a focusing lens with a focal length of 3 inches. The sensor integrates distance data with changes in reflectivity and angularity of the target. As the surface becomes rougher, the reflected light intensity drops significantly, but the general shape of the response curve remains constant with the maximum fixed at a given distance. If the same target is always being used in an application and the surface remains clean, the variations in the surface reflectivity are eliminated. However, if different surfaces are involved, a reference probe will be required to compensate for the reflectivity fluctuations. In addition, this sensor is sensitive to rotation of the reflected target. Rapid reduction in light intensity is more pronounced at large rotation angles. A typical error of +/- 3% has been reported for rotations of +/- 5 degrees about the normal to the liquid surface.

A major advantage of the reflective FO sensor is that no physical contact with the liquid surface is required for measurement. There are a few applications in which surface reflectivity can change due to surface contamination; as a consequence, a closed measuring system is required to avoid the adverse effect of contamination. The optical sensing occurs in a closed environment in which the sensor tracks the movement of a mechanical plunger based on the reflections at the mirror. By using an optical version of a linear voltage displacement transformer, contact with the part to be measured is required by the mechanical plunger, thereby eliminating all other problems.

8.2.2 Displacement Sensor Using Microbending Concept

Microbending of an optical fiber results in energy loss. The parameters that influence microbending loss include the fiber NA, core size, and core-to-cladding ratio and the periodicity of fiber deformation. When a periodic microbend is introduced in a fiber, mode coupling occurs between modes with different longitudinal propagation constants. This mode coupling is inversely proportional to the wavelength of periodic bending (Λ). The difference for adjacent modes is dependent on the NA, fiber core index (n_1), core radius (a), modal group label (m), and number of modal groups (M). Higher-order modes (M) are coupled with small values of Λ and vice versa. Microbending transducers are based on coupling core modes to radiated modes. Maximum sensitivity occurs when the higher-order core modes are made to radiate [2]. This situation can be mathematically approximated when $M = n$ and the expression for the wavelength of periodicity bending (Λ) can be written as

$$\Lambda = [a\, n_1/1.44\, NA] \qquad 8.3$$

It is evident from this equation that as the NA increases, the periodic perturbation spacing must decrease for maximum sensor sensitivity. As the core size decreases, the spacing must also decrease. Generally, high-NA optical fibers and small-core fibers guide light more strongly and require more severe bending to lose light intensity. For a given fiber diameter, increases in the core-to-cladding diameter ratio will increase microbending sensitivity. For a given displacement, a large core will have more bending and, therefore, more light leakage. The light loss is dependent on the microbending periodic perturbation (Λ), or the wavelength of periodic bending. As this period approaches the magnitude of the ray period of light propagating in the optical fiber, a strong resonance is possible with loss at its maximum value (see Figure 8–7). The primary peak corresponds to the periodicity that couples the highest-order modes to the radiant modes. As this period increases, the sensitivity decreases due to poor coupling efficiency with low-order modes. In cases of a higher incidence angle, which corresponds to higher-order modes, greater microbending loss is expected.

One important application of microbending sensors is displacement measurement. The transmission-versus-displacement response curve has three distinct regions. In the first region, the complaint coating absorbs the initial displacement movement that results in limited fiber bending. The small bending level results in the radiating of leaky modes. The second region yields a linear response over about 60% of the transmission range. This region is used for the sensing function. As the displacement increases further, light depletion occurs in the second-order modes, leading to a considerable reduction in sensor sensitivity [3].

Performance stability is the major consideration in the design of microbending sensors. Most optical fibers for this type of sensor are coated with a polymeric material as a protective mechanical buffer. Such coatings may flow under heavy thermal loads at elevated temperatures. The sensor accuracies are limited to no better than 1% due to potential hysteresis problems. However, use of high-performance metallized coatings may reduce this particular problem.

8.3 FO-Based Flow Sensors

Flow measurement is the most critical control parameter in a wide range of applications such as engine control, power generation, and industrial processes. Often the operating environment is severe. The flow sensor can be subjected to high electrical noise, explosive environments, very high temperature, radiation environments, and areas of difficult access. FO sensors have the ability to perform with and retain desired accuracy under these environmental conditions. Four basic concepts have been used in the design of FO-based flow sensors:

- Rotational frequency monitoring of turbine in the flow field.
- Differential pressure measurement across an orifice.
- Frequency monitoring of a vortex-shedding device.
- Laser Doppler velocimetry.

8.3.1 Turbine Flow Sensors

Turbine flowmeters require that they be put in the flow path, which causes flow obstruction. The flowmeter has moving parts. Both these features are highly undesirable, but the approach is accurate and repeatable and has broad industrial acceptance. Turbine flowmeters have a rotating device called a rotor, which is positioned in the flow stream as shown in Figure 8–8. The rotor velocity is proportional to the flow of fluid passing through the device. The rotor rotational speed is monitored using a reflective FO probe. As each vane or blade of the rotor passes the probe, a reflective optical pulse is generated. Since the sensor output is in a digital format, the problems associated with FO intensity-modulated analog signals are eliminated.

8.3.2 Differential Pressure Flow Sensors

In a differential pressure flow sensor, the volume of the fluid is forced to flow through a restricted area with reduced cross-section. The restriction causes an increase in the flow rate at that point. The net effect is that there is a pressure drop associated with the restriction. The change in pressure is proportional to the flow of fluid. The concept of a differential pressure flow sensor using FO technology is illustrated in Figure 8–5. The simplicity of differential pressure flow sensors gives them the potential for widespread application. These sensors are best suited for flow measurements involving low viscosity fluids and gases.

8.3.3 Flow Sensors Using Vortex-Shedding Concept

The concept of a vortex-shedding flow sensor is illustrated in Figure 8–8. As the fluid passes over a bluff body or object, alternating vortices are generated from each side of the bluff body. The vortex formation generates a pressure pulse. The frequency of pressure pulses is proportional to the fluid velocity. The pressure-pulse frequency is defined as

$$F_{pp} = [S\ V\ W] \qquad 8.4$$

where S is the Strouhal number (a dimensionless constant that is dependent on the fluid viscosity), V is the flow velocity, and W is the width of the bluff body or object. The unique feature of this sensor is that it is inherently digital. Therefore, only a pressure excursion, not the magnitude of the pulse, needs to be determined. Since the access is limited due to tight spacing, the reflective FO technique for monitoring diaphragm is the most logical choice for this application [4]. A sensor using optical fibers with core diameters close to 300 microns and NAs equal to 0.37 is expected to offer high accuracy. One distinct advantage of this approach is the very small size of the sensing element obstructing the flow path. A disadvantage is that it appears to have dead zones that are associated with natural resonant frequencies in the sensor response curve.

242 Chapter 8: Fiber Optic Sensors for Various Industrial Applications

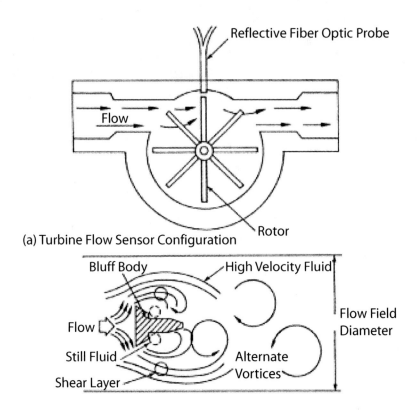

(a) Turbine Flow Sensor Configuration

(b) Turbine Flow Meter using Shedding, Vortex Concept

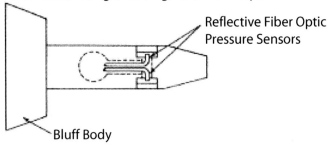

(c) Reflective Pressure Sensor for Shedding Vortex

Figure 8–8 *Various FO flow meter configurations. Courtesy of Instrument Society of America, Research Triangle, North Carolina 27709.*

8.3.4 Flow Sensor Using Laser Doppler Velocimetry (LDV) Concept

Laser doppler velocimetry (LDV) offers a noninvasive mechanism to measure fluid flow [5]. The dual beam concept provides two beams of equal intensity focused on a common point. The intersection of the two beams creates an interference with the fringes being generated. The fringe pattern is parallel to the bisector of the optical beams. As a particle passes through the fringe pattern (consisting of dark and light regions), light is scattered. The intensity across the laser beam associated with scattering is Gaussian, as illustrated in Figure 8–7. If the distance between the two fringes is D_f and the time to pass from one fringe to another is t, then the velocity component in a direction normal to the fringe is equal to V_x, which can be written as

$$V_x = D_f [F + 1/t] \qquad 8.5$$

where F is the frequency of fluctuation in the scattered light intensity, which is also known as the Doppler frequency. A polarization-maintaining optical fiber is required on the input side to maintain high measurement accuracy. A large core, MM, high-NA fiber must be used if maximum collection efficiency is the principle requirement. For optimum sensor performance, intensity fluctuations must be kept to a minimum.

8.4 FO Techniques for Chemical Analysis

FO techniques offer several distinct advantages. Chemical analysis can be done in situ in real time without compromising the authenticity of the results [5]. The sensing techniques do not disturb the chemical process under progress. The sample size can be extremely small, and the sensing locations can be in remote areas that are normally difficult to access. Three major disadvantages are sensitivity to ambient light, relatively slow response time due to the required reaction with various reagents, and shortened lifetime if a high radiation level is used to achieve high sensitivity. Four distinct techniques can be used for qualitative and quantitative chemical analysis: including fluorescence, scattering, absorption, and relative index change techniques.

A chemical-analysis technique can employ either a transmissive or a reflective FO sensor configuration as shown in Figure 8–9. In the transmissive sensor configuration, the light travels down a transmitting FO line (either a single fiber or a bundle of several fibers), passes through a gap that contains the sample to be analyzed, is captured in a receiving fiber, and transmits the light to a sensitive photodetector. In the case of the reflective sensor configuration, a bifurcated optical probe is used. The light travels down the transmitting section, reflects off the sample material, and is accepted in the receiving section that is attached to a

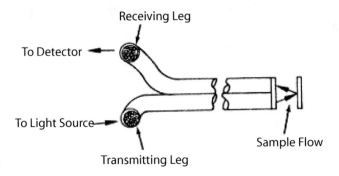

(a) Reflective System Configuration for Chemical Analysis

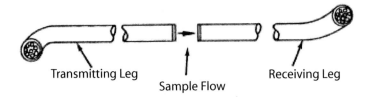

(b) Transmissive System Configuration for Chemical Analysis

Figure 8–9 *FO (a) reflective and (b) transmissive system configurations using fluorescence technique for chemical analysis.*

photodetector. Since the fluorescence technique is widely used for quantitative and qualitative chemical analysis, this technique will be described in greater detail.

8.4.1 FO Sensor Using Fluorescence Technique

The FO fluorescence technique used for chemical analysis is also known as remote fiber fluorimetry (RFF). In an RFF system, a high-intensity light source is coupled into a single quartz optical fiber with a large core diameter. The optical source light (see Figure 8–9) travels along the length of the fiber with minimum loss as a result of total internal reflection. Upon exiting the optical fiber, the rays of light impinge on the sample under chemical analysis, which gives off a characteristic fluorescent emission at a specific wavelength. The emission is detected by the same fiber and travels back to a sensitive photodetector that, with the help of a high-speed computer, can provide both the qualitative and quantitative chemical-analysis data with high accuracy and within a minimum test period.

The fluorescence can be improved by incorporating a sapphire microlens if it is absolutely essential.

Accurate measurements using the RFF technique have been obtained with chloride, iodine, uranyl, iron, plutonium, and sulfate ions. Ground-water contamination such as toluene and xylene, associated with gasoline or crude-oil spills, has been detected successfully by the RFF technique. Chemical analysis on aluminum and other metals can be performed using a reagent immobilized in the form of a powder and attached to a bifurcated FO probe [5]. The metal reacts with the reagent, yielding a fluorescent signal that can be detected with an accuracy of 0.027 ppm.

Partial oxygen pressure can be measured by a fluorescence technique using a fluorescent dye. When the dye is illuminated with a high-intensity blue light (340-to-700 nm spectral range) using an FO probe, the blue light impinges on the dye and gives off a characteristic green fluorescence. The level of fluorescence diminishes with the higher levels of oxygen that pass through the gas-permeable membrane, which reacts with the dye. The partial oxygen pressure is a function of the ratio of blue light intensity to green light intensity. The sensor provides accuracy better than 1% with a response time of less than 100 seconds over a pressure range of 0 to 150 torrs. Halide contaminations including iodine, bromide, and chloride can also be detected using the fluorescence technique with a detection accuracy of 2 ppm, 6 ppm, and 200 ppm, respectively [5]. This technique can be expanded to measure the pH value or the concentration of a metal ion.

8.4.2 Potential Applications for Chemical Analysis

Chemical analysis plays a key role in industrial process and medical monitoring to meet specific accuracy requirements. The principle advantage for both categories is accurate real-time measurements with minimum perturbations to the process being monitored. Remote chemical analysis is most desirable in a chemical plant because of personnel-safety reasons. Central analysis data monitoring using remote FO sensors must be given serious consideration over costly in-line conventional analysis equipment. Most materials do not fluoresce or possess absorption or scattering phenomena in the desired wavelength spectral region. This means that appropriate reagents need to be developed that are compatible with optical techniques for the majority of chemical materials and processes to be monitored with high accuracy.

The medical application of this sensor is important because of the extremely small-size sensor requirements for direct use in the human body. Fluorescence techniques are best suited for biological-process applications. For example, fluorescence can differentiate between the deceased tissue and normal arterial tissue or between a cancerous tissue and noncancerous tissue.

8.5 FO Sensors for Temperature Monitoring

Several operational considerations dictate the need for FO sensors for temperature monitoring. Sometimes sensors are required to monitor the temperature in strong electromagnetic field environments. Temperature monitoring sensors with metallic leads will experience large eddy currents in such environments, which will not only generate noise and excessive heat in the sensors but also cause errors in temperature measurement. FO temperature sensors do not use metallic leads and provide quick response. Since these sensors are less perturbing to the operating environments, they provide highest accuracy in temperature measurements and maximum reliability under harsh operating conditions because of the superior thermal and mechanical characteristics of optical fibers. Several temperature-sensing design concepts for the FO temperature sensors are available, such as reflective, microbending, intrinsic, phase modulated, and intensity-modulated concepts. Since the reflective sensor design concept is the least expensive and exhibits relatively higher monitoring accuracy and linear response, this particular temperature sensor will be described in greater detail.

8.5.1 FO Temperature Sensor Using Reflective Design Concept

Critical elements of an FO temperature sensor using the reflective concept are shown in Figure 8–10. This sensor uses a bimetallic sensing element attached as a transducer to a bifurcated reflective FO probe and is capable of the accurate monitoring of temperatures under harsh operating environments. The bimetallic element can be designed to provide a snap action at a given temperature and can move abruptly with respect to the probe tip, thereby resulting in a switching action at a preset temperature point. The element can be designed to move continuously, if required, and provide movement proportional to temperature. Typical sensor response in an analog mode is depicted in Figure 8–10. A digital indication of temperature can be achieved using differential thermal expansion involving two target materials such as alumina and silica, which have significantly different coefficients of thermal expansion. This causes reflected light to enter the output optical filter and the net temperature is a function of relative movement to each other.

Active sensing materials can be placed in the optical path of a reflective probe to achieve significantly improved sensing function. A variety of materials are available for use as a temperature-sensing element, including birefringent materials, semiconductor materials, inorganic/organic liquid crystals, and fluoresce materials. In the case of liquid crystals, the change in reflectivity due to temperature change is dependent on the light-source wavelength. But the intensity of the reflected light is a function of temperature. The temperature working range is extremely limited, but temperature accuracies greater than $0.1°C$ have been achieved. The liquid-crystal concept is best suited for monitoring biological processes.

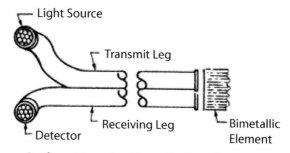

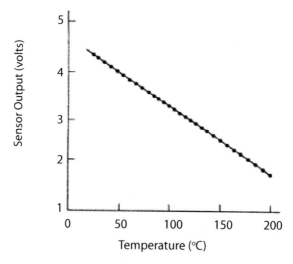

Figure 8–10 *(a) Sensor configuration (b) and response curve for a temperature sensor using bimetallic transducer.*

A reflective temperature-sensor design involves a birefringent crystal, which is an optically transparent crystalline material in which the refractive indices are different from the orthogonal polarized optical signals. The light, after passing through a polarizer, is reflected from a mirror and passes back through the crystal, polarizer, and optical detector. In these element materials, the birefringence is a strong function of temperature. The net change in the birefringence changes the received light intensity, which is proportional to the temperature being monitored. Again, this sensor type is limited to the temperature ranges of biological process.

8.6 FO Pressure Sensors

Pressure measurements are based on displacement measurements achieved by incorporating specific elements such as bellows and diaphragms. This means that the FO intensity-modulated concepts can be used for pressure measurements. These approaches may include transmission, reflection, and microbending phenomena. Regardless of the approach used in the design, the sensor configuration must offer compact size, minimum weight, high accuracy, and contact-free and EMI-free performance. The reflective diaphragm sensor is considered the optimum choice for pressure measurements because of the smallest size. Both the intensity-modulated and phase-modulated FO pressure-sensor designs are preferred for industrial applications. The interferometric design approach is widely used in the design of pressure sensors because the fiber itself can act as a pressure transducer capable of providing remarkable accuracy and wide frequency response. Such interferometric sensors are most attractive as acoustic-pressure measurement devices [5].

8.6.1 Pressure Sensors Using Transmission Approach

A transmissive sensor configuration along with sensor critical elements are shown in Figure 8–11. This figure illustrates a transmissive-sensor basic design in which a shutter interrupts the light path in a manner proportional to pressure intensity. Deployment of a reference channel and sensing channel provides the radiometric pressure data with accuracy greater than 0.1% over a pressure range up to 500 psi. The sensor sensitivity can be improved by placing two grating devices in the optical path: one fixed and the other movable. As the spacing between the grating devices becomes smaller, the sensitivity increases. This type of pressure sensor offers small size, precision, and a noncontact displacement pressure-monitoring device and is most ideal for monitoring diaphragm movement with great accuracy. The nonlinear response of a pressure-sensitive diaphragm can be corrected using compensating electronics, a sophisticated diaphragm design, and an optimum design of the FO probe. Implementation of these techniques will yield linear sensor response with very little hysteresis. Such a sensor will have a diaphragm linear movement close to 0.15 inches over a pressure range approaching 500 psi.

The sensitivity and linearity are the critical design parameters. An FO reflective probe is generally selected for integration in the design of a pressure sensor. In the case of a reflective FO sensor response curve, the front slope has generally better linearity and sensitivity compared to the back slope. Optimum design of the hemispherical FO probe is required to achieve high sensitivity and maximum linearity for the desired displacement range. Optimum probe design offers front-slope linearity better than 75% of the pressure operating range.

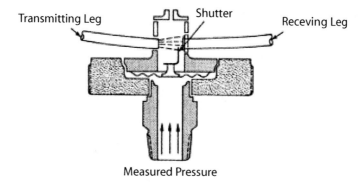

(a) FO-Based Pressure Sensor Using a Shutter for Modulation

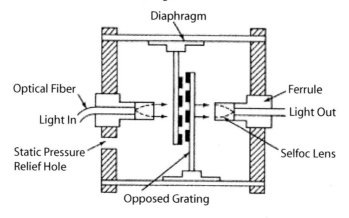

b) FO-based Pressure Sensor Using Moving Grating Device for Modulation

Figure 8–11 *FO pressure sensor using various modulation methods. Courtesy of Instrument Society of America, Research Triangle, North Carolina 27709.*

8.6.2 FO Pressure Sensor Using MEMS Technology

This particular sensor configuration involves a novel optically interrogated MEMS for the pressure sensor in which the entire MEMS structure is fabricated on an optical fiber (see Figure 8–12). The MEMS-based pressure sensor fabricated on an optical fiber has demonstrated a linear response to static pressure over the 0-to-80 psi range as illustrated in Figure 8–12. This sensor is best suited for applications where small-size dense arrays are of critical importance. The silicon diaphragm and the cavity-fiber interface act as reflectors forming a Fabry-Perot interferometer. Pressure causes the diaphragm to move, thereby changing the Febry-Perot reflectivity and allowing the measurement of the pressure parameter. This is because the changes in the

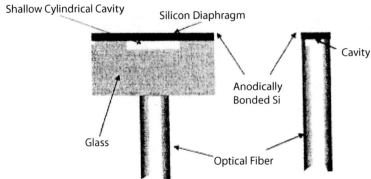

(a) Illustration of Two Sensor Configurations of an MEMS-Based Pressure Sensor Using Fabry-Perot Interferometry Concept

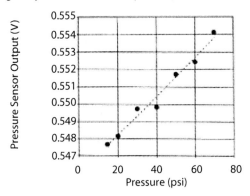

(b) MEMS-Based Pressure Sensor Output Versus Pressure

Figure 8–12 *(a) Sensor configuration and (b) sensor response for an MEMS-based pressure sensor using interferometry concept.*

amount of reflected light can be related to pressure. A linear response over various pressure ranges requires optimum values [6] of diaphragm thickness (7 mm), cylindrical cavity depth (0.64 mm), and cavity diameter (135 mm).

8.6.2.1 Design Requirements for MEMS-Based Pressure Sensor

The optical fibers used in the fabrication of the MEMS-based sensor are borosilicate MM fibers having refractive indices of 1.5098 at a wavelength of 850 nm. The sensor configuration shown in Figure 8–12 offers significant reduction in the size of the sensor and makes the packaging extremely simple and adhesive-free. The micromachining process can be used for the fabrication of an array of sensors which are capable of providing pressure maps with very impressive spatial resolution. The MEMS technology is most attractive for pressure-sensing applications because the small package size and precision sensing elements offer maximum economy and unlimited design flexibility in the selection of pressure range, bandwidth, and sensitivity. Furthermore, optical interrogation of the sensing elements allows the sensor to operate efficiently under harsh environments where conventional electronics are not able to operate. Optical interrogation is superior to electronic interrogation when the sensor is operating under high temperature, vibration, EMI, and dusty environments. The operating temperature of this particular sensor will be approximately that of the optical fiber. A unique design feature of this sensor is that the processing can be done on a commercial optical fiber as opposed to a wafer. The MEMS-based sensor design represents a state-of-the art device for pressure measurements with high accuracy.

The MEMS-based pressure-sensor configuration requires anodic bonding of a piece of an extremely thin silicon wafer onto the optical fiber end face over a cavity etched in the central portion of the fiber end face. This allows the device diameter to be the same as that of the optical fiber, which could be less than 100 microns. Due to small sensor size and the optical interrogation technique, multiplexed arrays of such sensors can be readily fabricated and are capable of providing pressure maps with exceptionally high spatial resolution. These arrays could play a critical role in future space-based systems and commercial and industrial applications such as automotive and medical applications.

8.7 FO-Based Magnetic- and Electric-Field Sensors

Frequent monitoring of current (magnetic field) and operating voltage are required for power utility companies and other industrial applications where high electrical power is consumed. FO-based sensors are best suited for monitoring current and voltages because they have inherent immunity against EMI and are relatively less expensive [7]. Various optical and optomechanical effects such as the Kerr effect, Faraday rotation effect, magnetostriction effect, and Pockels effect can be used in the design field-monitoring sensors. In the case of magnetic-field monitoring sensors, several intensity-modulation schemes using magneto-optical materials have been investigated to achieve the Faraday rotation effect. Microbending techniques using optical fibers can be used to design these sensors. Magnetostrictive materials have been used in a phase-modulation sensor configuration to measure magnetic fields. For electric-field sensing applications, intensity-modulation schemes using electro-optical materials are widely used. Phase-modulation schemes using piezoelectric fiber coatings are best suited for electric-field measurements where measurement accuracy is critical.

8.7.1 Magnetic-Field Sensor Using Intensity-Modulation Concept

A transmissive FO can be deployed to measure the magnetic-field magnitude and to determine the current if a magnetic-optical material is placed in the light path. The magneto-optical material can be incorporated in the optical fiber, or a bulk material can be inserted between two optical fibers in a transmissive mode, as illustrated in Figure 8–13. The magneto-optical material uses the Faraday rotation effect by producing a change in the refractive index proportional to the magnetic-field intensity. In the presence of a magnetic field, the plane of polarization is rotated, which directly modulates the intensity of the transmitted beam. The rotation of the polarized beam (θ) is proportional to the product of the magnetic field intensity (H) and the length of the magneto-optical material (L). The transmitted beam intensity is proportional to $\cos^2\theta$ [8].

The major advantage of incorporating the Faraday effect in the optical fiber is that the sensor can have a much longer light path compared to that available from a bulk optical material. The longer the light path, the higher the sensitivity of the sensor based on the Faraday effect will be. Sometimes the intrinsic fiber sensor may present a problem due to the interaction of the fiber, also known as residual birefringence, with polarization associated with the magnetic field. This residual birefringence in the optical fiber tends to reduce sensitivity. To maintain high sensitivity, SM optical fibers with low residual birefringence are recommended for these sensors.

FO transmissive sensors using Faraday-effect materials as sensing elements can measure currents in the range of 100 to 15,000 A with an accuracy of greater than 1%. The linearity of the sensor response is limited to a small portion of the dynamic range of the sensor. Metal-coated optical fibers using microbending effects are widely used to measure low currents with high accuracy. When the current is passed through the coating, the fiber is in the magnetic field, which can be monitored by the sensor. The magnetic-field intensity is directly proportional to the current flowing in the fiber. State-of-the art magnetic-field sensors offer very high sensitivity and are capable of detecting magnetic-field intensity as low as 5×10^{-12} Oe per meter.

8.7.2 Electric-Field Sensors Using Intensity-Modulation Technique

A transmissive FO sensor can be used to monitor the electric-field intensity [8] if an electro-optic material is placed in the light path. No component or additional material needs to be added to a glass optical fiber to achieve the intrinsic sensing. Electro-optic materials [8] are well known for exhibiting the Kerr or Pockels effects. In the case of a sensor operating on the Pockels effect, refractive-index change is directly proportional to the electric-field intensity. However, in the case

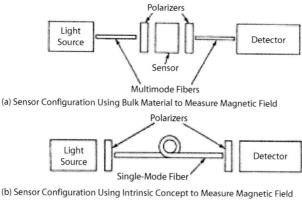
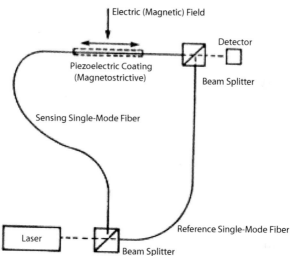

Figure 8–13 *FO sensor configurations to measure magnetic- and electric-field intensities.*

of a sensor operating on the Kerr effect, the refractive-index change is proportional to the square of the electric-field intensity. The Kerr effect is more prevalent in liquids than in solids. A transmissive electric-field sensor using $Bi_{12}GeO_{20}$ as an electro-optic material offers high sensitivity and linear output response over three orders of magnitude.

An electric-field or voltage sensor using a reflective probe is much more sensitive to the motion of the reflective target because a mirror is attached to a piezoelectric sensing element. When the reflective sensor is subjected to an electric field, the piezoelectric sensing element moves, causing the mirror to move and alter the reflected light in a manner proportional to the applied voltage. In summary, reflective FO sensors using electro-optic materials are best suited for applications where high sensitivity and excellent linearity over an ultrawide bandwidth are the principal requirements.

8.8 Summary

FO-based sensors best suited for commercial and industrial applications are described with particular emphasis on sensitivity and linearity over a wide dynamic range [5]. Immunity to EMI, RFI, and nuclear radiation makes FOs most attractive for industrial, military, and space applications. In addition, FO sensors are best suited for harsh operating environments. Performance requirements for process-control sensors widely used in the petrochemical and chemical industries; temperature sensors to monitor various industrial processes; displacement sensors to measure micromovements or tolerances in industrial, military and medical applications; pressure sensors to monitor absolute pressure in industrial processes and in steam power plants; and intensity-modulated sensors to measure electric-field and magnetic-field intensities are briefly summarized. Both transmissive- and reflective-type FO sensors are identified for various industrial applications. Critical requirements for FO transmission lines and probes are defined for optimum sensitivity. FO-core materials and NAs with minimum dispersion and insertion loss are identified for various industrial applications. Sensors using various sensing concepts involving reflective-surface change, refractive-index change, microbending, the laser Doppler velocimetry concept, MEMS technology, and the Mach-Zehender concept are briefly discussed with emphasis on accuracy, linearity, and sensitivity.

8.9 References

1. Rakucewicz, J. (1986). Fiber optic methods for level sensing. *Sensors,* 5–12.
2. Editor, (1982). Light loss due to microbending periodic perturbation. *Sensors,* 45–48.
3. Lugakos, N., et al. (1982). Microbend fiber optic sensoras extended hydrophone. *IEEE Journal of Quantum Electronics, QE-18*(10), 1633–1638.
4. Giallorenze, T., et al. (1982). Optical fiber sensor technology. *IEEE Journal of Quantum Electronics, QE-18*(4), 626–664.
5. Krohn, D. A. *Fiber optics sensor fundamentals and applications.* Research Triangle, NC: Instruments Society of America.
6. Abeysinghe, D. C. (2001). A novel MEMS pressure sensor fabricated on an optical fiber. *IEEE Photonics Spectral Letters, 13*(9), 993–995.

7. Harbner, R. E. (1985). *Fiber optic applications in electric power systems and measurement applications* (Part 2, IEEE Tutorial Course). Piscataway, NJ: IEEE.
8. Dandridge, A. (1980). Optical magnetic field sensors. *Electronic Letters, 16,* 408–409.

Index

A

Accuracy
    alignment 73, 110
    beam pointing 157
    tracking 115, 157
Active
    components 13, 180
    device 103, 180
Amplifier
    erbium-doped fiber 57, 75, 84, 90, 112, 117, 118, 119, 121, 137, 141, 142, 161, 165, 166, 167, 170, 172, 173, 174, 175, 184–192, 197
    gain profile 64, 187
    hybrid 32, 45, 141, 142, 156, 175
    optical 55, 56, 61, 75, 82, 90, 110, 112, 164–166, 167, 171, 172, 173, 184, 185, 190, 192, 198
    Raman 84, 156, 162, 167, 173, 175, 186, 188, 189, 191, 192, 198
Application
    aerospace 123, 133, 138
    commercial 123, 225, 226
    industrial 71, 123, 149, 229, 230, 236
    medical 200, 225–226, 229, 236, 245, 251, 254
    military 5, 11, 57, 58, 60, 73, 76, 80, 87, 107, 120, 230, 254
    remote sensing 59, 225
    space 5, 11, 57, 87, 120, 124, 254
Array
    detector 59, 201
    phased 87, 115, 117–118, 157, 160, 161, 198

B

Bandwidth
    modulation 95, 96
    spectral 26, 56, 81, 95, 138, 139
Beam
    beam forming 117–119, 160, 161, 162
    optical 62, 63, 75, 107, 110, 146
Biomedical applications
    angiogram 200
    catheter 200, 216–218, 221, 227
    dental surgery 200, 204–206, 227
    DNA 201, 202
    gynecology 200
    laser disc compression 200
    medical diagnosis 3, 200, 209, 216, 220
    ophthalmology 200, 216, 219, 225, 227
    percutaneous procedure 217
    photodynamic therapy (PDT) 73, 200, 212, 214, 227

Biomedical applications (cont'd)
   transmyocardial revascularization (TMR) 205, 217, 227

C

Coefficient
   absorption 7, 14, 184, 215, 221
   scattering 14, 215, 221
Communication
   dense-WDM 41, 55, 63, 77, 81, 102, 123, 156, 157, 164, 168, 174, 185, 198
   fiber optic 55–66, 156
   long-haul 3, 5, 13, 20, 24, 27, 84, 85, 98, 132, 139, 156, 162, 164, 184, 198
   optical 27, 41, 50, 58, 82, 84, 162, 168, 181
   wavelength-division-multiplexing (WDM) 5, 20, 22, 57–59, 60, 100, 106, 162, 163
Crosstalk
   amplitude 181
   channel 192, 193
   coherent 180, 181
   homowavelength 180, 181

D

Density
   aerosol 121
   atmospheric 111
Detector
   APD 157, 160, 184
   array 160, 201
   photon 55, 142, 182
   response 84, 184
   sensitivity 74, 184, 234

Device
   active 179
   electro-optic 3, 55, 82, 85, 93, 138, 201
   infrared 55, 85, 113
   passive 81, 178
   photonic 3, 69, 79, 80
Dispersion
   chromatic 6, 20–25, 26, 64, 76, 77, 78, 94–95, 100, 106, 165, 166, 168, 169, 173
   compensator 77, 85, 87, 98, 99, 170
   limits 23, 100
   slope 24, 99, 100, 121, 131, 165, 168
   zero 22, 24, 26, 168, 173, 176
DNA
   analysis 201–203
   argon laser 201, 204
   detector array 201
   electro-optic 201
   electropherogram 202
   fluorescence 202
   spectral identification 203
   spectrometer 150, 201

E

ECM
   programmable delay line 87, 88, 90
   range-gate-pull-off (RGPO) 87
   towed jammer 225
Effect
   absorption 20
   heating 217
Efficiency
   coupling 8, 9, 240
   laser 110, 213, 216

Index 259

quantum 184
Endoscopic procedure
    colon imaging 223
    endoscope 216, 223, 224
    field-of-view (FOV) 209, 224
    focused spots 224
    scanning element 223
    transverse motion 224
Error
    bias 105–106
    phase 134–135, 154
    wavefront 121
Expression
    coupling coefficient 27, 40, 53
    electric field 27, 28, 33
    group-delay 19, 27, 44, 95
    magnetic field 31

F

Fiber
    graded-index 4, 8, 148
    multi-mode (MM) 4, 10
    polarization-maintaining 25, 55, 65, 81, 113, 243
    single-mode (SM) 9
    step-index 4, 8, 21, 40, 75, 76
Fiber Bragg Grating (FBG)
    gain-flattening 190
    grating period 102, 171
    temperature compensation 153, 170
Fiber optic
    amplifier 165–167, 200, 203, 204, 207, 211, 212, 227
    Bragg gratings 87, 98, 123, 154, 156, 177, 179
    catheter 216
    communication 55–66
    control of T/R module 115, 157
    coupler 97

delay line 88
laser 71, 75, 76, 85, 87, 200, 203, 204, 207, 211, 212, 216, 219, 227, 229, 243, 246
link 159
ring-gyro 121
sensor 74, 144, 233
Filter
    add-drop 57–59, 80
    band-pass 57
    tunable 58, 59, 150, 151, 152, 209
    variable 59
Frequency
    cutoff 27, 28, 34, 36, 40, 47
    deviation 105
    modulation 105, 127
    normalized 20, 28, 32–33, 35–37, 39, 40, 44, 45, 46, 53
    optical 27, 32, 36, 79, 145, 169
Function
    Bessel 33, 37, 39, 40
    dyadic 27, 53
    Hankel 27, 28, 31, 33, 34, 45, 49, 53

G

Group
    delay 27, 33, 35, 44–47, 53, 95, 96, 98, 99, 101, 102–104, 119, 121, 127, 131
    delay spread 46, 47, 53

H

Hybrid
    amplifier 175
    technology 179, 224

I

Infrared
 countermeasures 113, 114
 detector 59
 flares 114
 sensor 55
 technology 114

J

Jamming
 directed IRCM 88, 90
 missile seeker 114
 shoulder-fired missile 113

K

Kerr effect
 bias error 253
Kerr effect, bias error 251, 253

L

Laser
 Er YAG 204, 205, 216, 225
 Fabry-Perot 59, 150
 Q-switch 71, 72, 107, 109, 121, 200, 221, 222, 223, 227
 solid-state 68, 74, 108, 109, 110, 114, 121, 175, 182, 183, 191, 192, 200, 204, 206, 213, 219, 222, 223, 226
 VCSEL 72, 182, 183
Life science
 cell specimen 208, 210, 223, 227, 228
 chromatography 208
 hyperspectral-imaging 210
 spectral information 208
 spectroscope 211

M

Material
 acrylate 4, 5, 8–12, 13, 14, 19, 20, 21, 21–25
 cladding 4, 9, 19, 32, 33, 37, 38, 39
 core 5, 9, 21, 23, 31, 32, 95, 170, 254
 fused silica 7, 13, 21, 32, 37, 93–95
 jacket 11, 41
 Kevlar 4, 11, 229, 230
 silicate glass 3, 4
 steel fibrous 4
 zinc chloride 4, 14
Mode
 dominant 36, 46, 50, 53
 free-space 38
 hybrid 27, 28, 30, 31, 32, 37, 38, 42, 43, 45, 46, 47, 48, 50, 52–53
 linearly polarized 35
 propogation 39, 40
 pseudo 35
Modulation
 bandwidth 95
 frequency 105
 intensity 105
 sinusoidal 105

N

Nonlinear
 effect 13, 14, 16, 22, 25, 166, 167–169, 174, 198
 response 248

O

Optical
 coupler 60, 90, 91, 97, 109, 152
 delay line 87, 88, 94

filter 55, 56, 57, 58, 167
gratings 24, 80, 130, 156, 178
isolator 61, 109
limiter 55, 56
power combiner 74, 76
switch 55, 65–69, 70, 71, 81, 85, 90–92, 93, 94, 129, 131

Optical amplifiers
bandwidth 164–165, 166, 167, 171, 184, 185, 192, 198
EDFA 57, 64, 85, 87, 156, 159, 163, 166, 167, 170, 171, 172, 173, 174–176, 184–188, 189, 192–196, 197
gain optimization 190
gain profile 64, 156, 187, 189, 190
noise figure 142, 173, 197, 198
Raman 84, 156, 162, 167, 173, 175, 186, 188, 189, 191, 192, 198
SOA 112, 172

Optical communication
channel capacity 27, 50, 168
dense-WDM 20, 57–59, 60, 68, 85, 100, 156
long-haul 5, 13, 20, 21, 24, 27, 84, 85, 98, 132, 139, 156, 162, 164, 166, 167, 173, 184, 198
metropolitan 173, 175, 198
wave-division-multiplexing (WDM) 174–177, 178, 185, 192, 198, 220

Optical sensors
detection of chemical agents 123, 138, 143, 144, 146, 154, 155
detection of toxic gases 143, 146
flow-metry 240, 241
industrial process control 230
interferometer 74, 79, 96, 103, 107, 132, 138, 139, 141, 229, 236, 249
pressure 229, 233, 248
quality control 146
surface reflectance 233, 234
vortex-shedding 240, 241

P

Polarization
fading 96, 121, 141
fluctuation 136
loss 25, 91

Power
density 15, 17, 20, 169, 197
fractional 32, 50, 51
modal 27, 28, 33, 48, 49, 50–53

Pump
efficiency 196
power 130–132, 167, 175, 186–187, 190–198
scheme 191, 192

R

Radar
missile 72, 117, 120, 160, 162
MMIC-based T/R modules 115, 157–159
phased array 3, 115, 117, 157, 160, 198
tracking 60, 115, 157, 198

S

Signal
bandwidth 99, 100, 109

Signal (cont'd)
    distortion-free 131
    signal-to-noise ratio 115, 126, 158
Spectral
    range 13, 14, 24, 25, 26, 55–56, 57, 58–59, 60, 65, 66, 71–72, 74, 78, 79, 81, 90, 95, 98, 105, 107, 121, 124, 131, 137, 138, 139, 152, 154, 156, 163, 167–169, 173, 174, 176, 181, 185–186, 187, 190, 192, 196, 197–198, 201, 202, 203, 205, 206–210, 211, 212, 214, 222, 225, 229, 245, 254
    response 41, 103
    width 13, 16, 134, 136, 139
Switch
    all-optical 66, 67, 68–70, 71, 86, 126, 129–132, 154
    cross-matrix 55, 65, 66
    ethernet 70
    laser-based 71
    MEMS-based 70
    speed 55
    ultrafast 126

T
Temperature
    birefringent 62
    compensated 102, 171
    dependence 100, 102, 170, 171
    fiber optic sensor 246

U
Ultraviolet, wavelength 220

V
Vortex 240, 241

W
Wavelength
    Bragg 99, 100, 102, 150, 151, 170, 171
    pump 131, 191, 192
    source 20, 88, 91, 94, 95, 96, 134, 136, 148, 215, 246

X
Xenon, short-arc lamp 212

Z
Zirconium, material 73